W9-BPR-915

8 Presenting the results   195

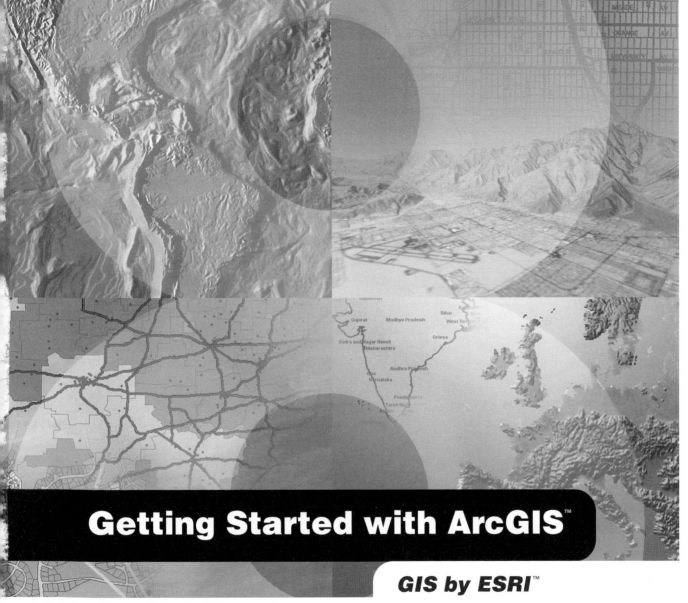

Getting Started with ArcGIS™

GIS by ESRI™

Bob Booth and Andy Mitchell

Contents

Getting to Know ArcGIS

Conducting a GIS Project

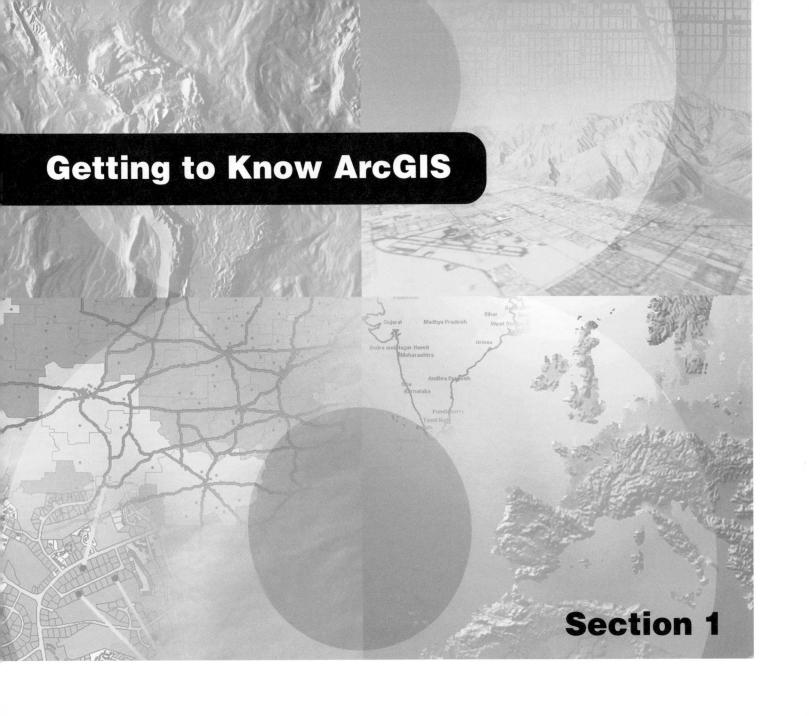

Getting to Know ArcGIS

Section 1

Introduction

Welcome to *Getting Started with ArcGIS*. This book is intended to help you get started using ESRI® ArcGIS™ software and to illustrate the methods and procedures involved in conducting a geographic information system (GIS) project. If you are new to GIS, this book is a great place to start—you can learn how to use a GIS to solve problems while you are learning to use ArcGIS.

This book is divided into two sections. The first section, 'Getting to Know ArcGIS', teaches you the basics of ArcGIS and GIS data. The second section, 'Conducting a GIS Project', begins with Chapter 4, 'Planning a GIS project', and is a sample GIS project that you can work through. The project is designed to let you work at your own pace, without the need of additional help. Readers who wish to complete the entire GIS project section of the book should plan to spend about eight hours of focused time on the project.

In order to get started, you will need ArcGIS installed on a Windows® machine. You will also need to install the ArcTutor tutorial data on your machine or on a networked drive. Proceed to Chapter 1, 'Welcome to ArcGIS', when you are ready to get started.

Welcome to ArcGIS

1

Welcome to ArcGIS, ESRI's premier GIS software. You can do virtually any GIS job at any scale of complexity with ArcGIS, from conducting a single analysis project on your own to implementing a vast, multiuser, enterprisewide GIS for your organization.

Use this book to learn what GIS is all about, and in just a short time you can begin to apply ArcGIS for all of your GIS needs.

Today, GIS is used by thousands of different organizations and hundreds of thousands of individuals to access and manage fantastically varied sets of geographically related information.

In this chapter, you will find samples of real-world uses of ArcGIS, a brief discussion of the different ways that GIS is used, some examples of how ArcGIS lets you use central GIS functions and, finally, some directions for learning more about ArcGIS.

What can you do with ArcGIS?

A tax assessor's office produces land use maps for appraisers and planners.

An engineering department monitors the condition of roads and bridges and produces planning maps for natural disasters.

A water department finds the valves to isolate a ruptured water main.

A transit department produces maps of bicycle paths for commuters.

A police department studies crime patterns to intelligently deploy its personnel and to monitor the effectiveness of neighborhood watch programs.

A wastewater department prioritizes areas for repairs after an earthquake.

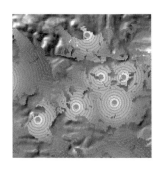

A telecommunication company studies the terrain to find locations for new cell phone antennae.

A hydrologist monitors water quality to protect public health.

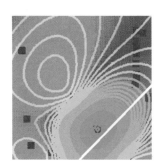

A pipeline company finds the least-cost path for a new pipeline.

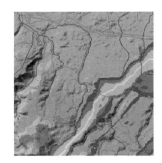

A biologist studies the impact of construction plans on a watershed.

An electric utility models its circuits to minimize power loss and to plan the placement of new devices.

A meteorologist issues warnings for counties in the path of a severe storm.

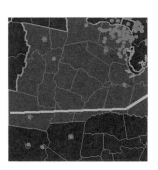

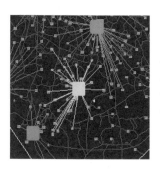

A business evaluates locations for new retail outlets by considering nearby concentrations of customers.

A police dispatcher finds the fastest route to an emergency.

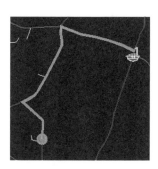

An emergency management agency plans relief facilities by modeling demand and accessibility.

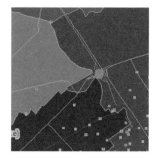

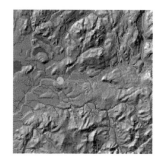

A water resource manager traces upstream to find the possible sources of a contaminant.

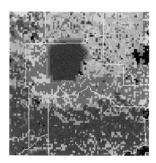

A fire fighting team predicts the spread of a forest fire using terrain and weather data.

Unique projects to daily business

You can use ArcGIS in different ways, depending on the complexity of your needs.

Some people use ArcGIS primarily as a single-user mapping and analysis tool, usually in the context of a well-defined, finite project. This common use of ArcGIS is sometimes called project GIS. Other people use ArcGIS in a multiuser system designed to serve an organization's ongoing needs for geographic information. Multiuser GIS is sometimes divided into departmental and enterprise GIS, according to a system's level of complexity and integration with the day-to-day operation of an organization.

This book presents ArcGIS in the context of project GIS because a project is a good, self-contained way to explore a variety of basic GIS functions.

Project GIS

In a GIS analysis project, an analyst faces a variety of tasks that can be grouped into four basic steps.

The first step is to convert a question, such as "Where is the best place for a new building?" or "How many potential customers are near this store?", into a GIS database design and an analysis plan. This involves breaking the question into logical parts, identifying what layers of data will be needed to answer each part, and developing a strategy for combining the answers to each part of the question into a final answer.

The next step is to create a database that contains the geographic data required to answer the question. This may involve digitizing existing maps, obtaining and translating electronic data from a variety of sources and formats,

making sure the layers are of adequate quality for the task, making sure the layers are in the same coordinate system and will overlay correctly, and adding items to the data to track analysis result values. Personal workspaces of file-based data and personal geodatabases are used to organize project GIS geodatabases.

The next step is to analyze the data. This usually involves overlaying different layers, querying attributes and feature locations to answer each logical part of the question, storing the answers to the logical parts of the question, and retrieving and combining those answers to provide a complete answer to the question.

The final step in a project-based analysis is to communicate the results of the analysis, usually to people who do not use GIS and who have different levels of experience in dealing with maps. Maps, reports, and graphs are all used, often together, to communicate the answer to the question.

Multiuser GIS

In a multiuser GIS, people in an organization—from a few in a single office to hundreds in different branches—use the GIS in different ways to support their daily tasks.

Departmental GIS refers to systems developed within a single department to support a key function of the department. For example, a planning department might routinely use GIS to notify property owners of proposed zoning changes near their property.

A departmental GIS is usually managed within the department and often has specialists devoted to different tasks. For example, a department might have its own system administrator, digitizer, and GIS analyst. Departmental GIS is often customized to automate and streamline procedures. For example, a planning department could use a GIS application that finds the names and addresses of parcel owners within a designated area and automatically generates notification letters.

An enterprise GIS spans departments in an organization. These large systems support multiple functions of an organization, from daily business to strategic planning. An enterprise GIS is usually managed as a part of the organization's information technology infrastructure. For example, a city's enterprise GIS integrates the business functions of building and maintaining the city. The engineering department builds the infrastructure for a subdivision using the same geodatabase that the planning department and assessor use to do their jobs.

An organization's entire network becomes the platform for an enterprise GIS. To provide access to many users, an enterprise GIS stores data in commercial relational database management systems (RDBMSs), such as Oracle®, Informix® Dynamic Server, and Microsoft® SQL Server™, that have been spatially enabled by ESRI's ArcSDE™ (formerly SDE®) software.

Using ArcSDE allows GIS data to be viewed and edited by many people simultaneously. To make the most of a networked system's capabilities, multiple seats of key applications, such as ArcCatalog™, ArcMap™, and ArcToolbox™, are deployed on desktop machines across an organization. Servers supply them with data and perform processor-intensive tasks.

The functions of a multiuser GIS are like those of a project GIS, but on a larger scale and operating in a continuous, cyclical fashion. Planning is crucial for multiuser systems, but the rewards—including increased operational efficiency, better allocation of scarce resources, consistency of information, and better-informed decisions—are tremendous.

Tasks you perform with ArcGIS

Whether you use GIS in a project or multiuser environment, you can use the three ArcGIS desktop applications—ArcCatalog, ArcMap, and ArcToolbox—to do your work.

ArcCatalog is the application for managing your spatial data holdings, for managing your database designs, and for recording and viewing metadata. ArcMap is used for all mapping and editing tasks, as well as for map-based analysis. ArcToolbox is used for data conversion and geoprocessing.

Using these three applications together, you can perform any GIS task, simple to advanced, including mapping, data management, geographic analysis, data editing, and geoprocessing.

ArcCatalog

ArcCatalog lets you find, preview, document, and organize geographic data and create sophisticated geodatabases to store that data.

ArcCatalog provides a framework for organizing large and diverse stores of GIS data.

Different views of your data help you quickly find what you need, whether it is in a file, personal geodatabase, or remote RDBMS served by ArcSDE.

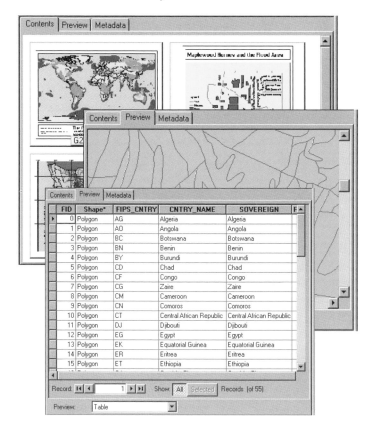

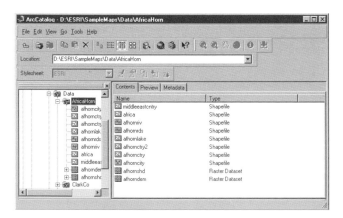

You can use ArcCatalog to organize folders and file-based data when you build project databases on your computer.

You can create personal geodatabases on your computer and use tools in ArcCatalog to create or import feature classes and tables.

You can also view and update metadata, allowing you to document your datasets and projects.

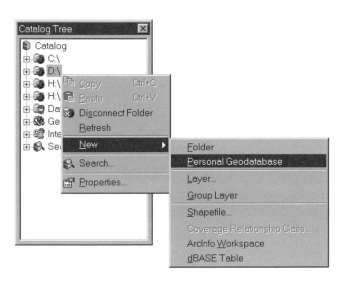

ArcMap

ArcMap lets you create and interact with maps. In ArcMap, you can view, edit, and analyze your geographic data.

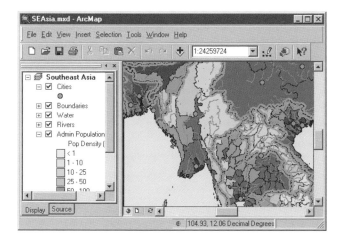

You can query your spatial data to find and understand relationships among geographic features.

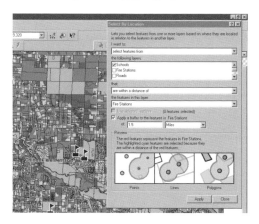

You can symbolize your data in a wide variety of ways.

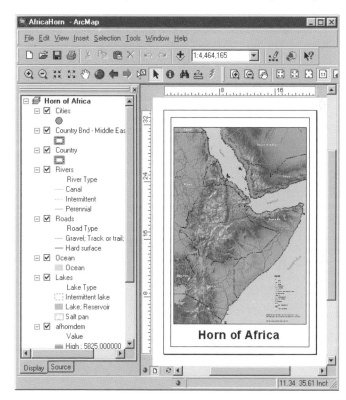

You can create charts and reports to communicate your understanding with others.

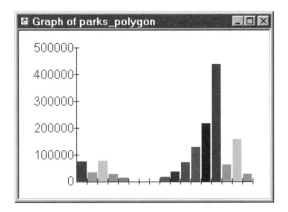

You can lay out your maps in a what-you-see-is-what-you-get layout view.

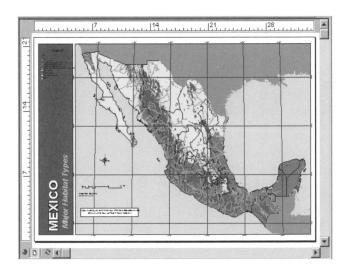

With ArcMap, you can create maps that integrate data in a wide variety of formats including shapefiles, coverages, tables, computer-aided drafting (CAD) drawings, images, grids, and triangulated irregular networks (TINs).

ArcToolbox

ArcToolbox is a simple application containing many GIS tools used for geoprocessing.

Simple geoprocessing tasks are accomplished through form-based tools.

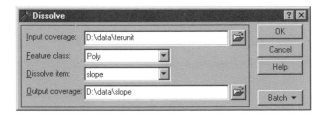

More complex operations can be done with the aid of wizards.

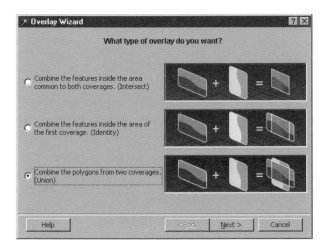

Accessing the ArcGIS desktop applications

The ArcGIS desktop applications can be accessed using three software products, each providing a higher level of functionality.

- ArcView® provides comprehensive mapping and analysis tools, along with simple editing and geoprocessing tools.

- ArcEditor™ includes the full functionality of ArcView, with the addition of advanced editing capabilities.

- ArcInfo™ extends the functionality of both to include advanced geoprocessing.

Note that there are two versions of ArcToolbox: the complete ArcToolbox, which comes with ArcInfo, and a lighter version of ArcToolbox, which comes with ArcView and ArcEditor.

ArcToolbox for ArcInfo comes with a complete, comprehensive set of tools (well over 150) for geoprocessing, data conversion, map sheet management, overlay analysis, map projection, and much more.

ArcToolbox for ArcView and ArcEditor contains more than 20 commonly used tools for data conversion and management.

You can use this book with ArcView, ArcEditor, or ArcInfo since it uses functionality common to all three software products.

See *What is ArcGIS?* for more information on ArcView, ArcEditor, and ArcInfo.

Tips on learning ArcGIS

This book is intended to help you learn the basics of ArcGIS. You can use the other books that come with ArcGIS to supplement the information in this book and to learn more about other tasks you can perform using ArcGIS.

When you want quick information about how to do a specific task, you can look it up in three handy reference books: *Using ArcCatalog, Using ArcMap,* and *Using ArcToolbox*. These books are organized around specific tasks. They provide answers in clear, concise steps with numbered graphics. Some of the chapters also contain background information if you want to find out more about the concepts behind a task.

Building a Geodatabase provides a step-by-step guide to building a geodatabase and implementing your geodatabase design in ArcGIS.

Two other books, *Modeling Our World* and *The ESRI Guide to GIS Analysis*, present the concepts behind GIS data models and geographic analysis, respectively.

The online Help system in ArcGIS also provides a wealth of information on using the software. Just click the Help button on any toolbar or dialog box. To get more information, see "Using this Help system" under the Help topic "Getting more help".

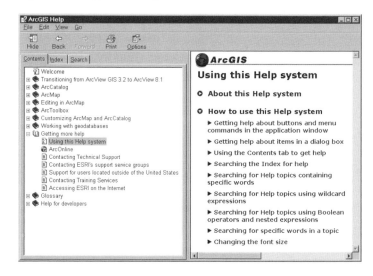

The "What's next?" section at the end of this book lists additional resources for learning ArcGIS and for getting help in completing your own GIS projects.

Exploring ArcCatalog and ArcMap

2

Maps are the most commonly used tools for understanding spatial information. Whether you do analysis or editing, produce wall maps or illustrate reports, design GIS databases or manage them—when you work with GIS you work with maps. ArcMap allows you to work with all of your geographic data in maps, regardless of the format or location of the underlying data. With ArcMap, you can assemble a map quickly from predefined layers, or you can add data from coverages, shapefiles, geodatabases, grids, TINs, images, and tables of coordinates or addresses.

Two other GIS applications—ArcCatalog and ArcToolbox—are designed to work with ArcMap. In ArcCatalog, you can browse, organize, and document your data and easily drag and drop it onto an existing map in ArcMap. Using the tools in ArcToolbox, you can project and convert data. If you are working in ArcInfo, ArcToolbox also has tools for sophisticated geoprocessing. It has never been easier to use the power of GIS.

In this chapter, you will create a map for a planning meeting of the Greenvalley City Council. You will use ArcCatalog to find the data and produce the map in ArcMap.

Introducing ArcCatalog

ArcCatalog is the tool for browsing, organizing, distributing, and documenting an organization's GIS data holdings.

In this exercise you work for the (fictitious) City of Greenvalley. The City Council is debating a proposal to build additional water mains downtown. As part of the process, the Council is reviewing water use in the downtown area.

You have been asked to make a map that shows the water mains in downtown Greenvalley and the relative water use at each parcel downtown.

To make the map easy to read, you will add the data to a general-purpose map of the town.

Start ArcCatalog

1. Click the Start button on the taskbar.

2. Point to Programs to display the Programs menu.

3. Point to ArcGIS.

4. Click ArcCatalog.

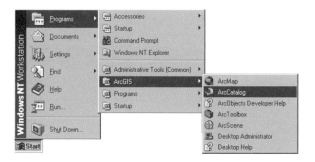

ArcCatalog starts, and you see two panels in the ArcCatalog window.

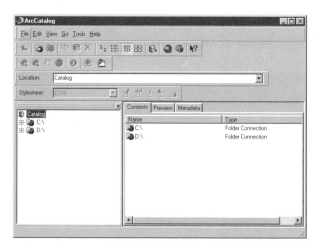

The Catalog tree on the left side of the ArcCatalog window is for browsing and organizing your GIS data. The contents of the current branch are displayed on the right side of the Catalog window.

Viewing data in ArcCatalog

When you need more information about a branch of the Catalog tree, you can use the Contents, Preview, and Metadata tabs to view your data in many different ways.

In this example, the ArcInfo coverage "cl" contains street centerlines. It is located on a computer's E:\ drive in a folder called City.

If you select a data source in the tree, you can view it in several ways, depending on the tab that you choose. Each tab has a toolbar associated with it that allows you to modify how you see your data.

These are Contents views:

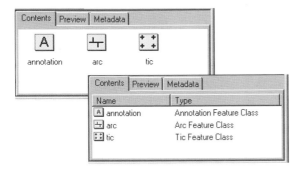

These are Preview views:

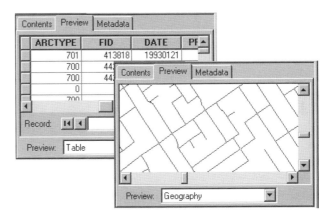

These are Metadata views:

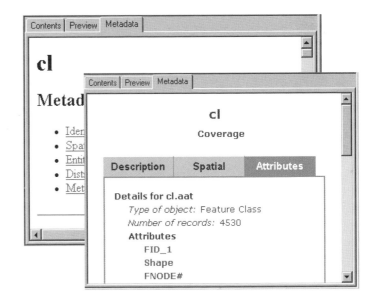

Connecting to your data

When you start ArcCatalog for the first time, the Catalog tree has a branch for each local hard drive. Branches for Coordinate Systems, Database Connections, Geocoding Services, Internet Servers, and Search Results can be added by clicking the Tools menu and clicking Options, then checking the check boxes next to the branches you want to add to the catalog. You can view the contents of a branch by double-clicking it or by clicking the plus sign beside it.

You can also create new branches in the Catalog tree to make it easier to navigate to your data. These branches are called connections.

Before continuing, you will need to know where the tutorial data has been installed on your system.

Make a connection to the tutorial data

Now you will add a connection to the folder that contains the tutorial data. This new branch in the Catalog tree will remain until you delete it.

1. Click the Connect to Folder button.

When you click the button, a window opens that lets you navigate to a folder on your computer or to a folder on another computer on your network.

2. Navigate to the ArcGIS\ArcTutor\Getting_Started\ Greenvalley folder on the drive where the tutorial data is installed. Click OK.

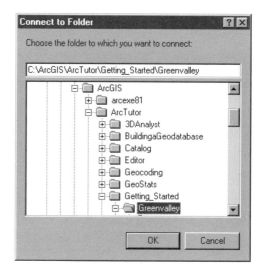

The new connection shows up as a branch in the Catalog tree.

Explore the Greenvalley folder connection

You can now look at the tutorial data that you have added.

1. Click the ArcGIS\ArcTutor\Greenvalley folder to view its contents on the right side of the ArcCatalog window.

2. Click the plus sign to expand the connection in the Catalog tree. This branch of the tree contains a folder, map documents, and a layer.

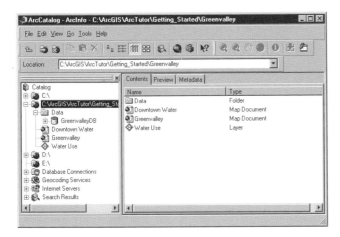

The Greenvalley folder has a special icon to show that it contains GIS data. By default, ArcCatalog recognizes many different file types as GIS data including shapefiles, coverages, raster images, TINs, geodatabases, projection files, and so on. If the list of recognized file types does not include a file type that you use in GIS analysis, you can customize ArcCatalog to recognize additional file types—for example, text files—as GIS data.

The Greenvalley map document is a general-purpose map of the City.

The Water Use layer shows a set of parcels in Greenvalley with a color scheme that indicates relative water use at each parcel.

Maps and layers

Maps and layers are important ways of organizing and displaying data in ArcGIS.

Maps, such as everyday paper maps, can contain many kinds of data. The data on a map is organized into layers, which are drawn on the map in a particular order. Each map contains a page layout where graphic elements, such as legends, North arrows, scale bars, text, and other graphics, are arranged. The layout shows the map page as it will be printed.

Layers define how a set of geographic features will be drawn when they are added to a map. They also act as shortcuts to the place where the data is actually stored—not necessarily the same place as where the layer file is stored. In this case, both the map and the layer refer to data that is stored in the Data folder.

If you store your geographic data in a central database, you can create maps and layers that refer to the database. This makes it easy to share maps and layers within an organization and eliminates the need to make duplicate copies of your data.

View the thumbnail sketch of the Greenvalley map

The right-hand panel of ArcCatalog displays datasets in many different ways. You can click an object in the left panel to view it in the right panel. One of the views that can be useful when you want to select a particular map is the thumbnail view.

1. Click the Thumbnails button on the Standard toolbar.

You can see the thumbnail sketch of the map.

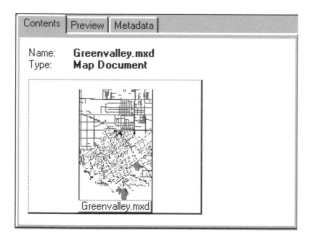

Open the Greenvalley map

You will use the Greenvalley map to provide context for the information that the City Council wants.

1. Double-click Greenvalley in the Catalog tree.

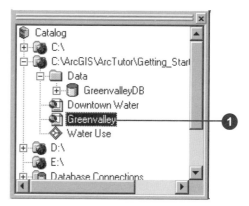

Double-clicking a map in the Catalog tree opens the map in ArcMap.

Sometimes you may want to start ArcMap without opening an existing map. You can start ArcMap by clicking the Launch ArcMap button in ArcCatalog.

Launch ArcMap button

You can also start ArcMap as you would any other program on your system, whether the Catalog tree is open or not.

Introducing ArcMap

ArcMap is the tool for creating, viewing, querying, editing, composing, and publishing maps.

Most maps present several types of information about an area at once. This map of Greenvalley contains three layers that show public buildings, streets, and parks.

You can see the layers in this map listed in the table of contents. Each layer has a check box that lets you turn it on or off.

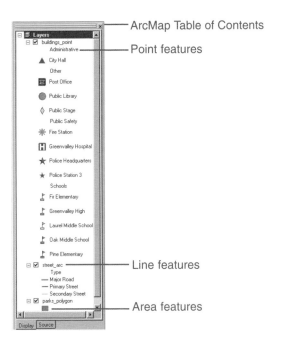

ArcMap Table of Contents

Point features

Line features

Area features

Within a layer, symbols are used to draw the *features*. In this case, buildings are represented by points, streets by lines, and parks by areas. Each layer contains two kinds of

information. The spatial information describes the location and shape of the geographic features. The attribute information tells you about other characteristics of the features.

In the park layer, all the features are drawn with a single green fill symbol. This single symbol lets you identify areas that are parks, but it does not tell you anything about the differences between the parks.

In the street layer, the features are drawn with different line symbols according to the type of street that the lines represent. This symbol scheme lets you differentiate streets from other types of features and tells you something about the differences between the features as well.

In the buildings layer, the features are drawn with different point symbols. The shapes and colors of the symbols allow you to differentiate the institutions that they represent. All of the schools are grouped together and drawn with a particular symbol, so you can easily differentiate schools from the hospital or from City Hall. Each school symbol is drawn with a different color, which lets you differentiate Pine Elementary from Greenvalley High.

Working with maps

ArcMap offers many ways to interact with maps.

Exploring

Maps let you see and interpret the spatial relationships among features. You could use the map you have just opened to find City Hall, to identify parks near schools, or to get the names of the streets around the library.

Analyzing

You can create new information and find hidden patterns by adding layers to a map. For example, if you added a layer of demographic information to the Greenvalley map, you might use the resulting map to define school districts or find potential customers. If you added layers of geology and surface slope, you might use the map to identify areas at risk for landslides.

Presenting results

ArcMap makes it easy to lay out your maps for printing, embedding in other documents, or electronic publishing. You can quickly make great maps of your data. When you save a map, all of your layout work, symbols, text, and graphics are preserved.

ArcMap includes a vast array of tools for creating and using maps. In the rest of this chapter, you will use some of these tools.

Customizing

Maps are tools for getting a job done. You can create maps that have exactly the tools you need to help you complete your job quickly. You can easily customize the ArcMap interface by adding tools to existing toolbars (or removing them) or by creating custom toolbars. You can save these changes to the interface with a particular map or for every map that you open.

You can also use the Visual Basic® for Applications (VBA) programming language included in ArcMap to create new tools and interfaces. For example, you can create a VBA tool to make a table of the addresses of houses in a selected area. Once the tool is created, you can associate it with a custom toolbar and save it with a map for anyone to use.

Programming

You can build completely new interfaces for interacting with your maps and create new, specialized classes of features. ArcGIS is built using Microsoft's Component Object Model (COM); all of the COM components are available to developers using a COM-compliant programming language. For more information about customizing ArcMap and ArcCatalog, refer to *Exploring ArcObjects.*

Exploring a map

You can explore a map in several ways. The Tools toolbar contains frequently used tools that let you navigate around the map, find features, and get information about them.

Zoom in and get information

If you want to see an area of the map in greater detail, you can zoom in to the map.

1. Click the Zoom In button.

2. Drag a box around one of the parks to zoom in to it.

When you drag a box on the map after clicking the Zoom In button, the map zooms to the new area. You can click the Back button to jump back to the previous map extent.

3. Click the Identify Features button and click the park.

When you click a feature with the Identify Features tool, the Identify Results window appears. You can inspect the attributes of the feature from this window.

If the tool finds several features where you clicked, it lists each feature on the left side of the window. You can click the features in this list to view their attributes on the right side of the window.

4. Close the Identify Results window.

Zoom to the map's full extent

If you have zoomed in to the map and want to see all of it, you can quickly zoom out to the map's full extent.

1. Click the Full Extent button.

Now you can see the full extent of the map. The map scale is around 1:95,000 (depending on your screen setup and the size of the ArcMap window), which you can see on the Standard toolbar. (If the map scale is not around 1:95,000 change it by clicking in the text box, replacing the text with 1:95:000, and pressing Enter.)

At this scale, the building symbols are not visible. The Maximum Visible Scale *property* of this layer has been set to 1:70,000. You will change some of the properties of a layer later in this chapter.

Find a feature

The Find button lets you search a map for features that match your search criteria. The area you want to map is around the Greenvalley City Hall, so you will find City Hall and zoom to it.

1. Click the Find button.

When you click the Find button, the Find dialog box appears. You can search for features from a particular layer or from all layers on the map.

2. Type "City Hall" in the Find text box. Click the In layers dropdown arrow and click buildings_point. Click In fields, then click the dropdown arrow, and click NAME. Click Find.

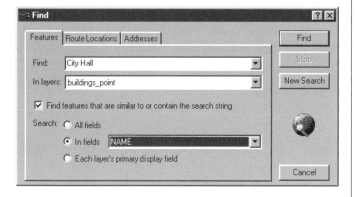

City Hall appears in the list of features that the tool has found.

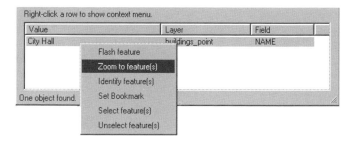

3. Right-click City Hall and click Zoom to feature(s).

The map zooms to the City Hall. As the scale is now greater than the 1:70,000 threshold, the building features appear on the map, and you can see the blue triangle symbol for City Hall.

4. Click Cancel to close the Find dialog box.

The map now shows some of the area that you need to map for the City Council.

When you chose Zoom to feature(s), another option on the list was Set Bookmark. A spatial bookmark preserves a particular map extent so that you can zoom back to it whenever you want.

Spatial bookmarks are saved with a map, so anyone who opens a map can quickly zoom to a particular bookmarked area.

Zoom to a bookmarked area

Because you use this map to provide a context for other information, you have created some spatial bookmarks for the areas you frequently map. Downtown Greenvalley is one of these areas.

1. Click View and point to Bookmarks.

2. Click Downtown Greenvalley.

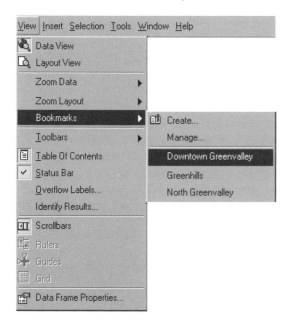

Now the map is zoomed to the downtown area. This map extent and scale has been used for previous maps of downtown Greenvalley. The map you are making will be easy for the Council members to compare with the other maps of the downtown area.

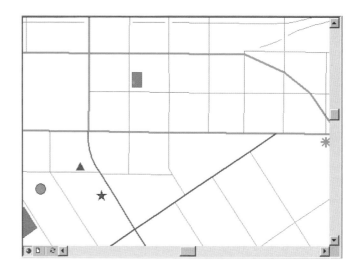

ArcMap provides an excellent interface for interactively exploring existing maps. You can use the tools you have just used and others to answer questions about particular features, find features, and view your maps at a variety of scales.

You can change the information that is displayed on maps by adding and removing layers and changing the way that layers are displayed.

In the next part of this chapter, you will add data to your map and change the properties of a layer.

Adding a layer to a map

Now that you have opened a map of Greenvalley and set the extent to downtown, it is time to make the map you need. The City Council wants the map to include downtown water use and the location and size of existing water mains. You will start by adding the Water Use layer to your map.

1. Position the ArcMap and ArcCatalog windows so that you can see both of them.

2. Click the Water Use layer in ArcCatalog and drag it onto the map. You can click and drag any layer from the ArcCatalog tree onto an open map in ArcMap.

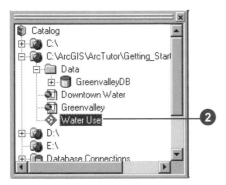

The layer shows parcels drawn with a graduated color ramp. Just as the roads and buildings were drawn with predetermined symbols when you opened the Greenvalley map, this layer is drawn with a particular set of symbols.

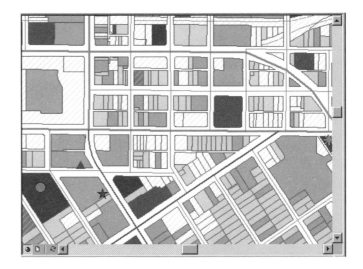

A layer serves as a shortcut to data. It also tells ArcMap how the data should be drawn. You can store layers in a place that is accessible to everyone in your organization who needs a particular set of data; the data will be displayed the same way for each of them.

As useful as layers are, sometimes they are not available. Fortunately, you can add raw geographic data to a map just as easily as you can add a layer.

Adding features from a database

When you add features directly from a coverage, shapefile, or database, they are all drawn with a single symbol.

Now you will add the water main features to your map.

1. Position the ArcMap and ArcCatalog windows so that you can see both of them.

2. Click the plus sign next to the Data folder in the Catalog tree to view the contents of the folder.

3. Click the plus sign next to GreenvalleyDB. GreenvalleyDB is a geodatabase that contains the remainder of the data you will be using. The data in this geodatabase is organized in five feature datasets: Hydrology, Parks, Public Buildings, Public Utility, and Transportation.

4. Click the plus sign next to Public Utility.

5. Click watermains_arc and drag it onto your map.

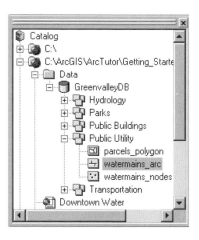

Watermains_arc is a feature class—a collection of features represented with the same geometry (shape). In this case, the features are polyline shapes that represent the pipes in the water distribution system.

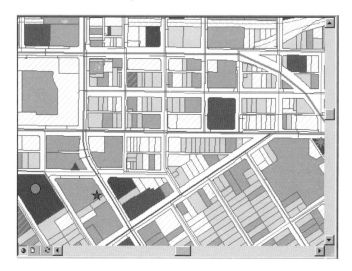

Geodatabases containing feature datasets and feature classes are how ArcGIS applications manage geographic information. In Chapter 3, you will learn more about these and other GIS data types.

Changing the way features are drawn

The Council wants to know the approximate sizes of the water mains downtown, so you must assign some new symbols to the features.

1. Right-click watermains_arc in the ArcMap table of contents and click Properties.

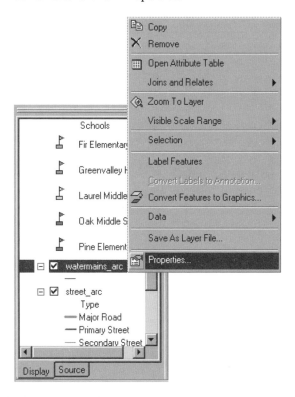

The layer Properties dialog box appears. You can use this dialog box to inspect and change a wide variety of layer properties.

The water mains feature class includes several attributes of the water mains. As the Council wants to know the sizes of the water mains, you will group the mains into five classes based on their diameter attribute.

2. Click the Symbology tab on the Properties dialog box.

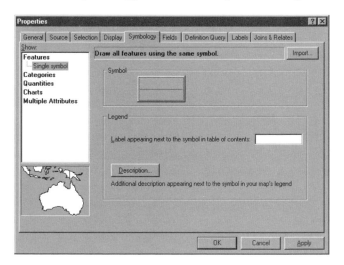

You can change the symbol scheme for the layer, as well as its appearance in the table of contents, from this tab.

3. Click Quantities. The panel changes to give you controls for drawing with graduated colors.

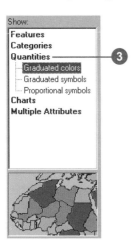

4. Click Graduated symbols. The panel changes to give you controls for drawing with graduated symbols.

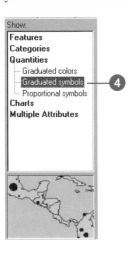

5. Click the Value dropdown arrow and click DIAMETER. ArcMap assigns the data to five classes using the Natural Breaks classification (Jenks' method).

Now the width of the line symbols indicates the diameter of the water mains. You want the water mains to be blue, so you will change the base symbol.

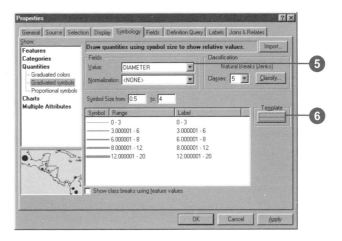

6. Click Template.

When you click Template, the Symbol Selector dialog box appears. Here you can choose predefined symbols, such as the Highway line symbol, or you can design your own symbols.

7. Click Color. The color selector dialog box appears. You can select one of the predefined colors from this palette or click More Colors to mix your own colors using one of several popular color models.

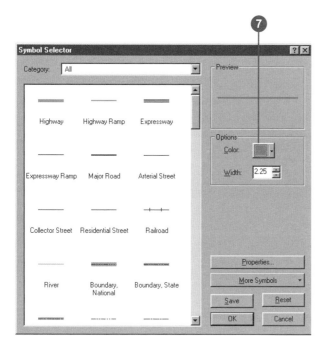

8. Choose a dark shade of blue and click OK.

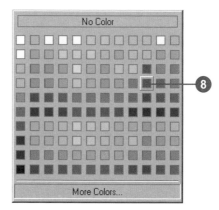

Now all of the water mains will be drawn with dark blue lines, with the line width representing the diameter of the water main.

9. Click OK on the Properties dialog box to see your map with the new line symbols.

As you have seen, ArcMap has a rich set of line symbol selection and editing tools. These and other tools also work with point and polygon symbols.

Once you have set the symbolization for a layer to your satisfaction, you can preserve it for later use by saving the map (later in this chapter) or by saving the layer as its own layer file such as the Water Use layer you added (see *Using ArcMap* for step-by-step instructions).

Adding labels to a map

The map now shows some of the street centerlines and water mains with similar symbols. To avoid confusing a map reader, you will add street names on the map and change the street centerline symbol.

1. Right-click street_arc in the table of contents.

2. Click Label Features.

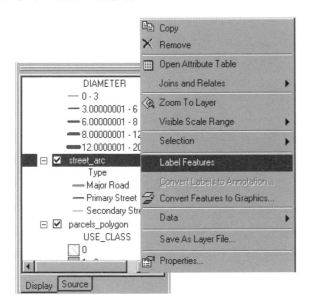

ArcMap adds the names of the streets to the map.

Change the street centerline symbol

1. Right-click street_arc in the table of contents again and click Properties.

2. Click the Symbology tab.

3. Click Features, then click Single symbol.

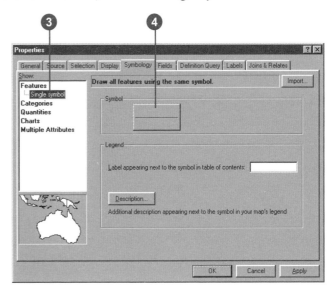

The street centerlines will now be drawn with a single symbol. You will change the default line color to a light gray, so the centerlines will be visible but unobtrusive.

4. Click the Symbol button.

The Symbol Selector appears.

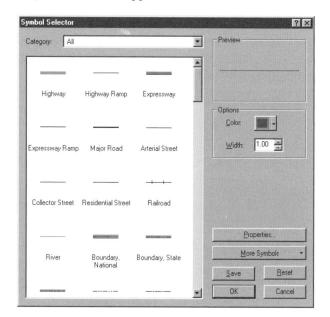

5. Click Color. Click a light gray and Click OK.

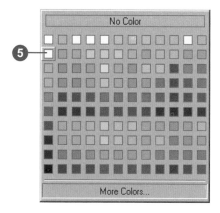

6. Click OK on the Properties dialog box.

Now the centerlines will be drawn in a light gray, so they will not be confused with the water mains.

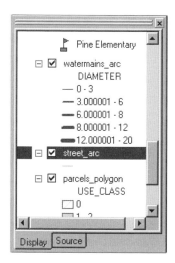

Working with the map layout

All of the data you need is now on the map and has symbols. The map that you are making for the Council meeting will be printed in color on an 8.5" x 11" sheet of paper and distributed to each council member.

1. Click View and click Layout View.

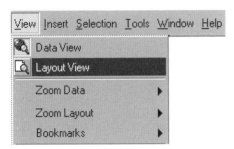

Now you can see the map on a virtual page. The layers of data appear in a *data frame* on the page. Data frames are a way of organizing the layers you want to see together on a map.

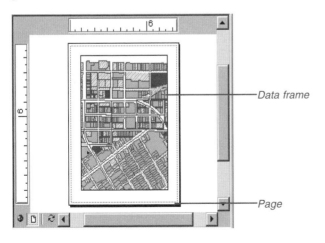

There is always at least one data frame on a map. This one is called Layers; you can see its name at the top of the ArcMap table of contents.

You can add additional data frames to a map to compare two areas side by side or to show overviews or detailed insets.

You can see all of the data frames on your map in Layout view. If you switch back to Data view, you will see the layers that are in the active data frame. The active data frame is shown in boldface type in the table of contents.

In Layout view you can change the shape and position of the data frame on the page, add other map elements such as scale bars and legends to the map, and change the page size and orientation.

The Layout toolbar is added to the ArcMap interface when you choose Layout View.

You can use the tools on the Layout toolbar to change the size and position of the virtual page on your screen or to zoom in or out of the virtual page.

You can also use tools from the Tools toolbar in Layout View to change the extent of the layers that are shown in the data frame.

2. Right-click on the page and choose Page Setup.

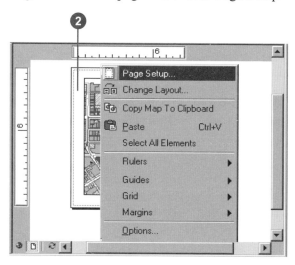

3. Click Landscape under Map Size and under Printer Setup to change the page orientation then click OK.

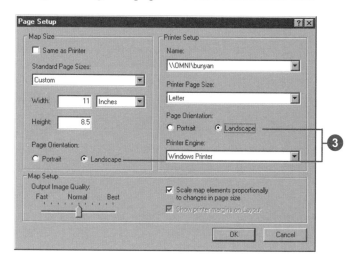

Now the page is in landscape orientation.

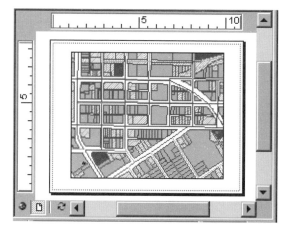

You will add a scale bar, North arrow, legend, and title to the page to help the Council members use the map.

First, you will make some space on the page for these other map elements by reducing the size of the data frame.

4. Click the Select Elements button.

5. Click the data frame to select it. The data frame is now outlined with a dashed line and has selection handles at its corners and edges.

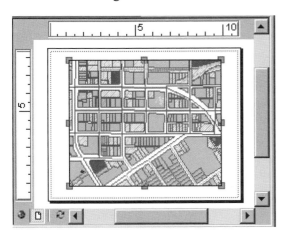

6. Point to the selection handle at the lower-right corner of the data frame. The cursor becomes a two-pointed resize cursor. Click the corner and drag it up and to the left.

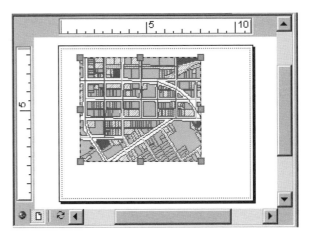

Add a scale bar

1. On the Insert menu click Scale Bar.

The Scalebar Selector dialog box appears.

2. Click one of the scale bars and click OK.

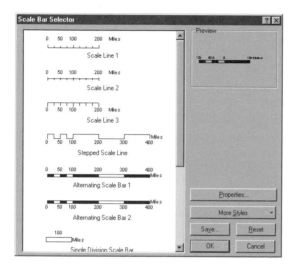

3. Click the scale bar and drag it to the empty space below the left side of the data frame.

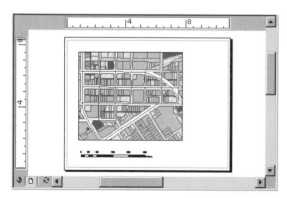

Add a North arrow

1. On the Insert menu click North Arrow.

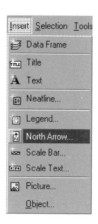

The North Arrow Selector appears.

2. Click one of the North arrows and click OK.

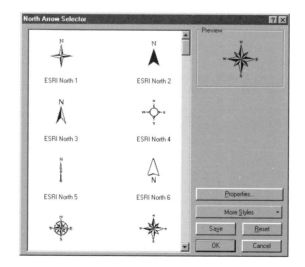

3. Click the North arrow and drag it to the empty space below the data frame and to the right of the scale bar.

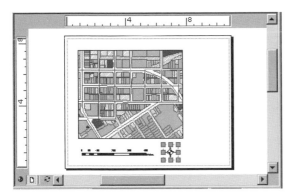

Add a legend

1. On the Insert menu click Legend.

The Legend Wizard appears.

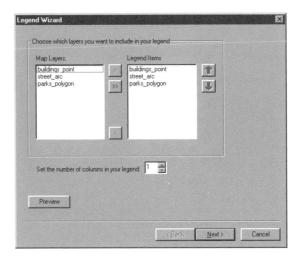

Changing Legend Wizard parameters alters the appearance of the legend on your map. The Legend Wizard takes you through five dialog boxes that allow you to change the layers included in your legend, the appearance of the legend title, the appearance of the legend frame, the size and shape of the symbol patches used to represent line and polygon features, and the spacing between legend items.

In this case, the default legend parameters are appropriate for your map. However, legend parameters can be modified at any time by right-clicking the legend in the layout view and choosing Properties from the menu that appears.

2. Click Next several times to step through the wizard, accepting the default legend parameters. Click Finish when done.

The legend appears on your map.

3. Click the legend and drag it to the empty space to the right of the data frame.

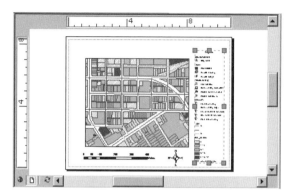

You can click on the blue selection handles to resize the legend so that it fits along the right side of the page.

Add a title

1. On the Insert menu click Title.

A partial title, "Greenvalley", appears in the layout view.

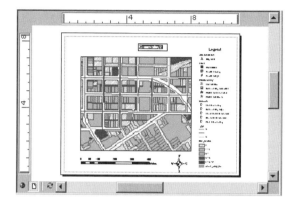

Greenvalley is the name of the map document, but you will need a more explanatory title on the map.

2. Type "Downtown Greenvalley Water Mains and Water Use". Press Enter, then click and drag the title to the top and center it on the page.

Saving a map

You have made a lot of changes to this map. Because you want to keep the new map that you have created and also keep the old template map, you will use Save As to save this map under a new name.

1. Click File and click Save As.

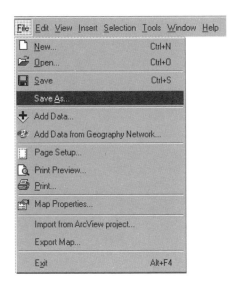

2. Navigate to the Greenvalley folder.

3. Type "Downtown Water". Click Save.

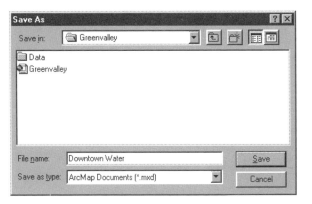

Now you are ready to print a copy of this map for the City Council.

Printing a map

You can easily print the maps you have composed in ArcMap. The Layout view lets you arrange map elements, such as data frames, scale bars, and North arrows, on the page exactly as you want them to print.

You can print your maps using any printer on your network, and you can choose to print using Windows, PostScript®, or ArcPress™ (if installed) printer engines.

1. Click File and click Print.

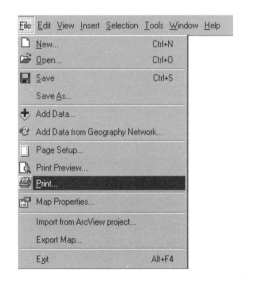

The Print dialog box appears. You can change the default printer by clicking Setup.

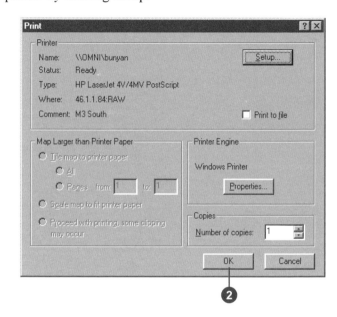

2. Click OK.

You are ready to take the map to the Council meeting.

Close ArcMap and ArcCatalog.

3. Click File and click Exit, or simply click the Close button (x) in the upper-right corner of the ArcMap window. Do the same for ArcCatalog.

What's next?

While making your first map, you learned how to start and use two GIS applications: ArcCatalog and ArcMap.

In the next chapter you will learn more about GIS data and how to work with various data types. Working in the field of GIS analysis inevitably means working with geographic data in a variety of different formats. Understanding the advantages and limitations of each format is an important first step in any project and is essential to the project you will begin in Chapter 4, 'Planning a GIS project'.

Exploring GIS data

IN THIS CHAPTER

- **Geographic data models**

- **Coverages**

- **Shapefiles**

- **Geodatabases**

In Chapter 2, 'Exploring ArcCatalog and ArcMap', you worked with a map and layers. The layers on a map are based on GIS data. When you added the water mains to the map, you added data from a feature class stored in a geodatabase. Other formats for GIS data include shapefiles, coverages, and rasters. GIS data formats vary, but they all store spatial and attribute information.

Much data has a spatial component that may not be immediately apparent. For example, customer databases often include addresses. With a suitable street dataset, these addresses can be plotted as points or geocoded. Similarly, tables of sales figures can be linked by a query statement to a feature class of sales territories and displayed on a map.

It is useful to understand the different GIS data types and database models when you are conducting an analysis project. This chapter contains a brief introduction to common types of GIS data and database models.

Geographic data models

ArcGIS stores and manages geographic data in a number of formats. The three basic data models that ArcGIS uses are vector, raster, and TIN. You can also import tabular data into a GIS.

Vector models

One way of representing geographic phenomena is with points, lines, and polygons. This kind of representation of the world is generically called a *vector* data model. Vector models are particularly useful for representing and storing discrete features such as buildings, pipes, or parcel boundaries.

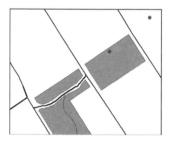

Points are pairs of x,y coordinates. Lines are sets of coordinates that define a shape. Polygons are sets of coordinates defining boundaries that enclose areas.

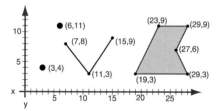

Coordinates are most often pairs (x,y) or triplets (x,y,z, where z represents a value such as elevation).

The coordinate values depend on the geographic coordinate system the data is stored in. Coordinate systems are discussed in more detail in Chapter 6, 'Preparing data for analysis'.

ArcGIS stores vector data in feature classes and collections of topologically related feature classes. The attributes associated with the features are stored in data tables.

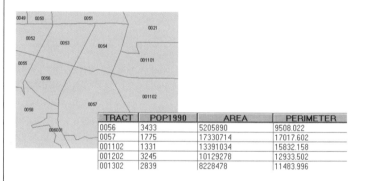

TRACT	POP1990	AREA	PERIMETER
0056	3433	5205890	9508.022
0057	1775	17330714	17017.602
001102	1331	13391034	15832.158
001202	3245	10129278	12933.502
001302	2839	8228478	11483.996

ArcGIS uses three different implementations of the vector model to represent feature data: coverages, shapefiles, and geodatabases.

Raster models

In a *raster* model, the world is represented as a surface that is divided into a regular grid of cells.

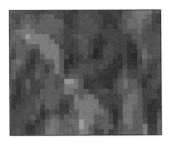

The x,y coordinates of at least one corner of the raster are known, so it can be located in geographic space.

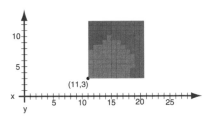

Raster models are useful for storing and analyzing data that is continuous across an area. Each cell contains a value that can represent membership in a class or category, a measurement, or an interpreted value.

Raster data includes images and grids. Images, such as an aerial photograph, a satellite image, or a scanned map, are often used for generating GIS data.

Grids represent derived data and are often used for analysis and modeling. They can be created from sample points, such as for a surface of chemical concentrations in the soil, or based on classification of an image, such as for a land cover grid. Grids can also be created by converting vector data.

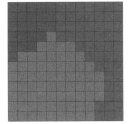

Grids can store continuous values, such as for an elevation surface.

They can also store categories, such as for a grid of vegetation types.

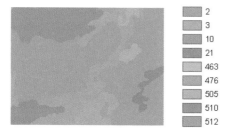

	2
	3
	10
	21
	463
	476
	505
	510
	512

Grids storing categorical information can store additional attributes about each category. For example, a grid of vegetation types might store—for each category—a numeric code, the name of the vegetation type, a habitat suitability rating for certain wildlife species, and a general type code. This is unlike feature data, where attributes are stored for each individual feature.

Value	Count	Name	Suitability	Type
2	30672	Cropland and pastureland	4	Agriculture
3	3339	Urban and industrial	5	Urban
10	212	Clearings and brushfields	5	Cleared
21	1383	Cottonwood	4	Riparian
463	142	Ash-Cottonwood	3	Woodland
476	7205	Oak	3	Woodland
505	1112	Douglas fir	2	Forest
510	6557	Mixed evergreen-broadleaf	3	Forest
512	7943	Douglas fir-Hemlock-Cedar	1	Forest

The smaller the cell size for the raster layer, the higher the resolution and the more detailed the map. However, because the cells form a regular grid over the whole surface, decreasing the cell size to store higher resolution data substantially increases the total volume of data that must be stored.

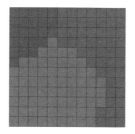

ArcGIS can recognize and use rasters from image files in many different file types and from grids stored in workspaces. You can add raster datasets to a map just as you would features, and you can inspect and organize them with ArcCatalog.

TIN models

In a *triangulated irregular network* model, the world is represented as a network of linked triangles drawn between irregularly spaced points with x, y, and z values. TINs are an efficient way to store and analyze surfaces.

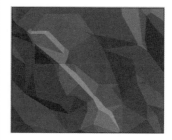

Heterogeneous surfaces that vary sharply in some areas and less in others can be modeled more accurately, in a given volume of data, with a triangulated surface than with a raster. That is because many points can be placed where the surface is highly variable, and fewer points can be placed where the surface is less variable. ArcGIS stores triangulated surfaces as TIN datasets. As with rasters, you can add TIN datasets to a map in ArcMap and manage them with ArcCatalog.

For more information about raster data and TINs, see *Modeling Our World: The ESRI Guide to Geodatabase Design.*

Tabular data

You can think of a GIS as a database that understands geometry. Like other databases, ArcGIS lets you link tables of data together. Just about any table of data can be joined to an existing feature class or raster dataset if they share an attribute. For example, you may have a shapefile of census tracts with a tract number field and a tabular file of additional census data also containing a tract number field. You can link the census data to the shapefile's attribute table and map the additional data.

Geolocating is another means of getting tabular data on a map. Perhaps the simplest example of geolocating is plotting points based on tables of geographic coordinates. For example, you can plot the locations of soil samples based on latitude–longitude values obtained from a global positioning system (GPS) receiver. You can also plot points by geolocating tables of addresses on an existing street network. This is often called address geocoding.

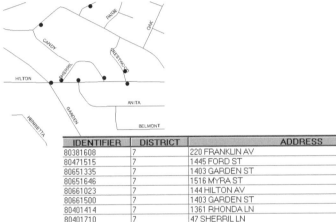

IDENTIFIER	DISTRICT	ADDRESS
80381608	7	220 FRANKLIN AV
80471515	7	1445 FORD ST
80651335	7	1403 GARDEN ST
80651646	7	1516 MYRA ST
80661023	7	144 HILTON AV
80661500	7	1403 GARDEN ST
80401414	7	1361 RHONDA LN
80401710	7	47 SHERRIL LN

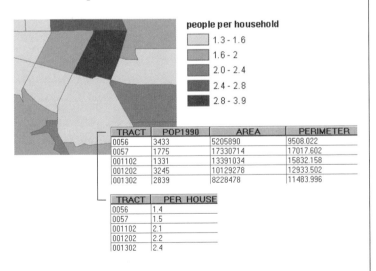

people per household

- 1.3 - 1.6
- 1.6 - 2
- 2.0 - 2.4
- 2.4 - 2.8
- 2.8 - 3.9

TRACT	POP1990	AREA	PERIMETER
0056	3433	5205890	9508.022
0057	1775	17330714	17017.602
001102	1331	13391034	15832.158
001202	3245	10129278	12933.502
001302	2839	8228478	11483.996

TRACT	PER HOUSE
0056	1.4
0057	1.5
001102	2.1
001202	2.2
001302	2.4

Formats of feature data

ArcGIS supports both file-based feature models and database management system (DBMS) feature models.

The two file-based models are coverages and shapefiles. Coverages and shapefiles employ a *georelational data model*. They store the vector data for the features in binary files and use unique identifiers to link features to attributes stored in feature attribute tables in other files.

The DBMS feature model supported by ArcGIS is the *geodatabase data model*. In this model, features are stored as rows in a relational database table. The rows in the table contain both the coordinates and the attribute information for the features.

Coverages

Coverages are the traditional format for complex geoprocessing, building high-quality geographic datasets, and sophisticated spatial analysis.

Coverages contain primary, composite, and secondary feature types. The *primary* features in coverages are label points, arcs, and polygons. The *composite* features—routes/sections and regions—are built from these primary feature types.

Coverages may also contain *secondary* features: tics, links, and annotation. Tics and links do not represent geographic objects but are used to manage coverages. Annotation is used to provide text about geographic features on maps.

Primary features in coverages

Label points can represent individual point features, for example, wells. In the diagram below, the point in the upper left represents well number 57.

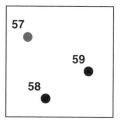

Label points also link attributes to polygons. Each polygon in a coverage has a single label point with its feature ID number, usually located near the polygon's center. The diagram below shows the label points of polygons 102 and 103.

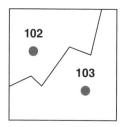

Arcs are connected sets of line segments, with nodes at the endpoints. A single arc can stand alone such as a fault line on a geologic map; several arcs can be organized into line networks such as stream or utility networks.

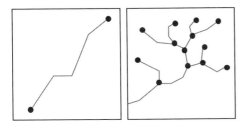

Arcs can also be organized into polygons that represent areas such as soil types.

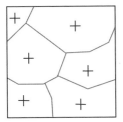

Nodes are the endpoints of and connections between arcs.

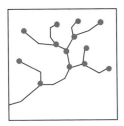

Nodes can have attributes, so they can represent point features in a network such as valves in a network of water mains.

Nodes are important for tracking how features in coverages are connected to each other; this is called *topology*. You'll learn more about coverage topology later in this section.

Polygons represent areas. They are bounded by arcs, including arcs that define island polygons. Polygons in a coverage may share arcs such as B and C below, but they do not overlap. Each point in an area falls within exactly one polygon so, for example, a point within polygon A is outside of polygon B.

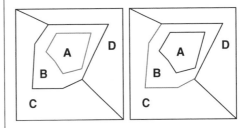

Composite features in coverages

Routes and *sections* are linear features that are composed of arcs and parts of arcs. Routes define paths along an existing linear network such as the route from a house to an airport along a street network.

Because points of interest on a network are not always at nodes, sections identify partial arcs. They record how far along a given arc a route begins or ends.

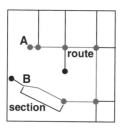

Regions are area features that are composed of polygons. Unlike polygons, regions can be discontinuous. For example, the mainland and an island can be mapped as two polygons, but they can belong to the same region.

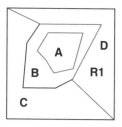

In the diagram above, polygons A and D belong to region R1.

Regions in a coverage can also overlap. For example, in a coverage of forest polygons, two regions that represent different forest fires could overlap if an area that burned in one year also burned in another.

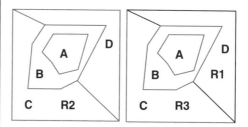

In the diagram above, region R2 and region R3 share polygon C.

Secondary features in coverages

Annotation features are text strings that describe a feature when a map is displayed or printed. Annotation can be positioned at a point, between two points, or along a series of points. Annotation is used to make maps easier to read and understand.

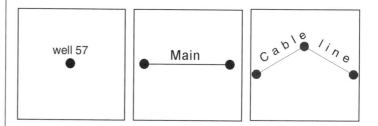

Annotation is stored in geographic coordinates, so it maintains its position and scale relative to the other coverage features when displayed.

Tics are geographic control points. They represent known locations on the ground and are used to register and transform the coordinates of a coverage.

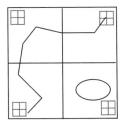

Tics allow features digitized from a paper map to be accurately transformed from digitizer units such as centimeters or inches to real-world units such as kilometers or miles. It is a good practice to use the same tic locations when you digitize sets of features from a map into different coverages so they will overlay correctly.

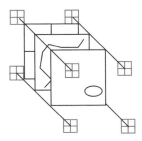

Links are displacement vectors that are used to adjust the shape of coverages, for example, to match the edges of adjacent coverages. Links consist of a from-point and a to-point.

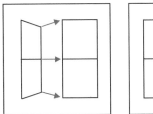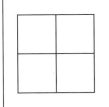

Coverage topology

Topology is the procedure for explicitly defining and using the spatial relationships inherent in feature geometry. The three major topological relationships that coverages maintain are connectivity, area definition, and contiguity.

Coverages implement topology and explicitly record these spatial relationships in special files. Storing connectivity makes coverages useful for modeling and tracing flows in linear networks. Storing information about area definition and contiguity makes it possible to find or merge adjacent polygons and to combine geographic features from different coverages with overlay operations.

Coverages store *connectivity* by recording the nodes that mark the endpoints of arcs. Arcs that share a node are connected. This is called arc–node topology. Each arc is a connected set of vertices with a from-node and a to-node.

The illustration below shows three arcs labeled 1, 2, and 3. Arc 1 starts from node 10 and goes to node 20. Its shape is defined by vertices a, b, c, and d. Arc 2 is connected to arc 1 at nodes 10 and 20.

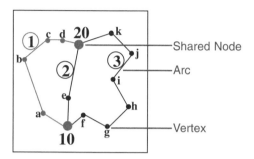

Coverages *define areas* by keeping a list of connected arcs that form the boundaries of each polygon. This is called polygon–arc topology.

In this illustration polygon A is defined by arcs 1 and 2.

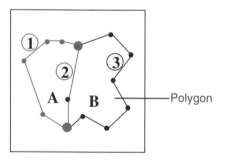

Coverages store *contiguity* by keeping a list of the polygons on the left and right side of each arc. This is called left–right topology. Polygons that share an arc are contiguous. In this illustration polygons A and B are contiguous because A is to the left of arc 2 and B is to the right.

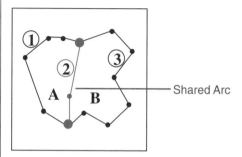

Storage of coverages

Coverages are stored in workspaces. A workspace is a folder in the file system. The workspace folder contains a folder named info and folders named for each coverage in the workspace.

Here, the workspace is called a_workspace, and the coverages are called a_coverage and b_coverage.

A coverage folder contains a set of files that store information about the features in the coverage (coordinates, topology, and so on). The attributes of coverage features are stored in feature attribute tables that are managed by an

INFO™ database. The info folder contains INFO data files and table definitions for each coverage.

In ArcCatalog, you see a coverage workspace as a GIS data folder. You can identify the geometry of a coverage (point, line, polygon, and so on) by its icon. You can also see the feature classes within a coverage.

Workspace in ArcCatalog

Here, you can see that a_workspace contains two coverages: a_coverage and b_coverage. The a_coverage contains an arc feature class and a tic feature class. This coverage has polygon topology, so it contains a polygon feature class and a label feature class as well. The dataset b_coverage is a line coverage, so it just contains arc and tic feature classes.

You may also see additional data tables in a coverage workspace if there are other tables stored in the INFO database, such as linked data tables or symbol lookup tables.

Shapefiles

Shapefiles are useful for mapmaking and some kinds of analysis. A great deal of geographic data is available in shapefile format.

Shapefiles are simpler than coverages because they do not store full topological associations among different features and feature classes. Each shapefile stores features belonging to a single feature class.

Features in shapefiles

Shapefiles have two types of point features: points and multipoints. They have line features that can be simple lines or multipart polylines. They also have area features that are simple or multipart areas called polygons.

Point shapes are simply single-point features such as wells or monuments. Here, well number 57 is selected.

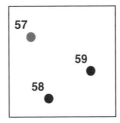

Multipoint shapes are collections of points that all represent one feature. A group of small islands could be represented as a single multipoint shape. Here, multipoint feature 22 is selected.

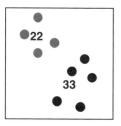

Line shapes can be simple continuous lines such as a fault line on a map. They can also be polylines that branch such as a river. Line shapes can also have discontinuous parts.

Polygon shapes can be simple areas such as a single island. They can also be multipart areas such as several islands that constitute a single state.

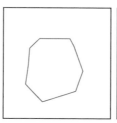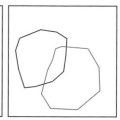

Polygon shapes can overlap, but the shapefile does not store topological relationships between them. The market areas of two stores could be represented as overlapping polygon shapes.

Storage of shapefiles

Shapefiles are stored in folders. A shapefile consists of a set of files of vector data in the shapefile and a dBASE® .dbf file containing the attributes of the features. Each constituent file shares the shapefile name.

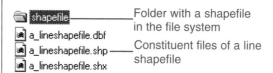

Folder with a shapefile in the file system

Constituent files of a line shapefile

A shapefile contains shapes of only one geometry: point, multipoint, line, or polygon.

When you look at a folder of shapefiles in ArcCatalog, you see the shapefiles as standalone feature classes.

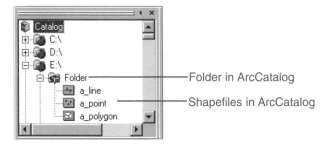

Folder in ArcCatalog

Shapefiles in ArcCatalog

Geodatabases

Geodatabases implement an object-based GIS data model—the geodatabase data model. A geodatabase stores each feature as a row in a table. The vector shape of the feature is stored in the table's shape field, with the feature attributes in other fields. Each table stores a feature class.

In addition to features, geodatabases can also store rasters, data tables, and references to other tables. Geodatabases are repositories that can hold all of your spatial data in one location. They are like adding coverages, shapefiles, and rasters into a DBMS. However, they also add important new capabilities over file-based data models.

Some advantages of a geodatabase are that features in geodatabases can have built-in behavior; geodatabase features are completely stored in a single database; and large geodatabase feature classes can be stored seamlessly, not tiled.

In addition to generic features, such as points, lines, and areas, you can create custom features such as transformers, pipes, and parcels. Custom features can have special behavior to better represent real-world objects. You can use this behavior to support sophisticated modeling of networks, data entry error prevention, custom rendering of features, and custom forms for inspecting or entering attributes of features.

Features in geodatabases

Because you can create your own custom objects, the number of potential feature classes is unlimited. The basic geometries (shapes) for geodatabase feature classes are points, multipoints, network junctions, lines, network edges, and polygons. You can also create features with new geometries.

All point, line, and polygon feature classes can

- Be multipart (for example, like multipoint shapes or regions in a coverage).

- Have x,y; x,y,z; or x,y,z,m coordinates (m-coordinates store distance measurement values such as the distance to each milepost along a highway).

- Be stored as continuous layers instead of tiled.

Point and *multipoint* geodatabase features are similar to the corresponding feature types in shapefiles.

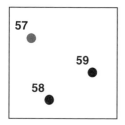
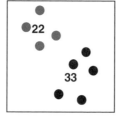

Generic point features could, for example, represent building locations in a city.

Custom point features could also represent buildings, but they might include an interface that would list the owner, area, and assessed value of the building or display a photograph or schematic of the building.

Network junction features are points that play a topological role in a network, somewhat like nodes in a coverage. There are simple and complex network junction features.

A *simple junction feature* might be used to represent a fitting that connects two pipes. It could have validation behavior that would ensure that connected pipes are of the correct diameter and materials.

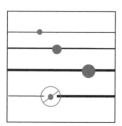

A *complex junction feature* plays a more complex role in a network. It can contain internal parts that play a logical and topological role in the larger network.

For example, a complex junction feature could be used to represent a switch in an electric power network. In one position the switch could connect point A to point B, while in another position it could connect point A to point C.

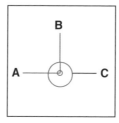
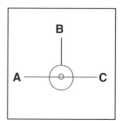

The switch might have editing validation rules that would control the types of power lines that could be connected to it. It could also have custom behavior that would draw the switch with different symbols depending on its state (open or closed, for example).

Line features are lines built from three kinds of segments: line segments, circular arcs, and Bézier splines. A single line could be built from all three parts, as in the illustration on the right, below.

Lines can be used to represent linear geographic features such as roads or contour lines. Line features can have custom drawing behavior that generalizes the line depending on the map scale or that controls the placement of annotation along the line.

Network edge features are lines that play a topological role in a network. They can be used for tracing and flow analysis.

Here, the network between A and B has been traced. The network contains simple and complex network edge features.

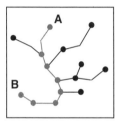

A *simple edge feature* is a linear network feature that connects to junction features at its endpoints. In this respect, simple edge features are similar to arcs, which have nodes at their endpoints. A simple edge feature could be used to represent a pipe in a water network.

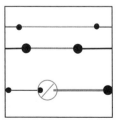

Simple edge features can have connectivity rules; for instance, a 10-cm pipe must connect to a 10-cm fitting. They can also have special class methods, so a pipe feature could calculate the pressure drop of a liquid flowing from one end to another based on the pipe diameter, roughness, and length. They can have special query, editing, and data entry interfaces.

A *complex edge feature* is a linear network feature that can support one or more junctions along its length, yet remain a single feature. In the example below, the line from A to B is a single complex edge feature.

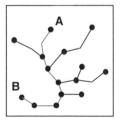

A power line could be represented as a complex edge feature. It could have junction features at its ends and additional junctions where other lines attach to it along its length. Like simple edge features, complex edge features can have special class methods and interfaces.

Polygon features represent areas. Their boundaries can be composed of line, circular arc, and Bézier spline segments—the same geometries used to create line features. They can be simple closed shapes, or they can have discontinuous parts. Polygon features can also have nested islands and lakes.

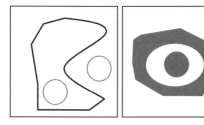

You can use polygon features to represent geographic features such as buildings, census blocks, or forest stands. As with other geodatabase features, polygon features can have customized behavior and interfaces. A custom building polygon could be drawn as a plan view at one scale, as a generalized building footprint at another, and with a point symbol at another. It could also have a custom interface for editing and viewing its attributes.

You can create your own custom geodatabases from scratch, or you can modify elements of an existing geodatabase. For more information about designing geodatabases and creating custom features, see *Modeling Our World: The ESRI Guide to Geodatabase Design* and *Building a Geodatabase.*

Topology in a geodatabase

Topology in a geodatabase allows you to represent shared geometry between features within a features class and between different feature classes. You can organize the features in a geodatabase to create planar topologies or geometric networks.

Feature classes can share geometry with other feature classes in a planar topology. For example, you might define a topological relationship between streets, blocks, block groups, and census tracts. The street segments define the boundary of the block they enclose. Groups of blocks can be collected into block groups, and block groups into tracts.

A planar topology is composed of a set of nodes, edges, and faces. When you update the boundary of one feature, the shared boundaries are updated as well.

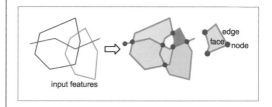

Topologically related edge and junction features within a dataset can be bound into a geometric network. This is useful when the features must be connected to each other with no gaps. For example, you could organize pipes, valves, pumps, and feeders into a water network.

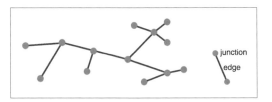

Storage of geodatabase features

Geodatabase features reside in geodatabases. A multiuser, versioned geodatabase can be implemented using ArcSDE software in any of the leading commercial all-relational databases. Single-user (or personal) geodatabases are implemented in a Microsoft Access .mdb file.

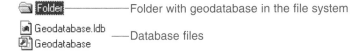

You access the database through ArcGIS applications including ArcMap and ArcCatalog.

Geodatabase feature classes each contain one geometric feature type. Related feature classes can be organized into feature datasets. Feature datasets are useful for organizing feature classes with a shared topology. They can also be used to organize feature classes thematically. For example, you might have three feature classes in a waterbodies feature dataset: points, representing ponds; lines, representing rivers; and polygons, representing lakes.

When you look at a geodatabase in ArcCatalog, you see the database tables as collections of feature datasets and feature classes, or simply as standalone feature classes.

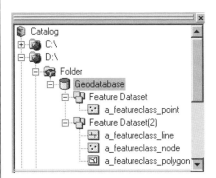

Geodatabase feature classes are stored with spatial indexes, so you can work efficiently with small areas of very large seamless databases. This eliminates the need to divide large, complex datasets into separate tiles.

Getting more information

There are many facets to each vector data format and many issues to consider when choosing one over another within a specific database design. For a full discussion of these issues, see *Modeling Our World: The ESRI Guide to Geodatabase Design* and *Building a Geodatabase*.

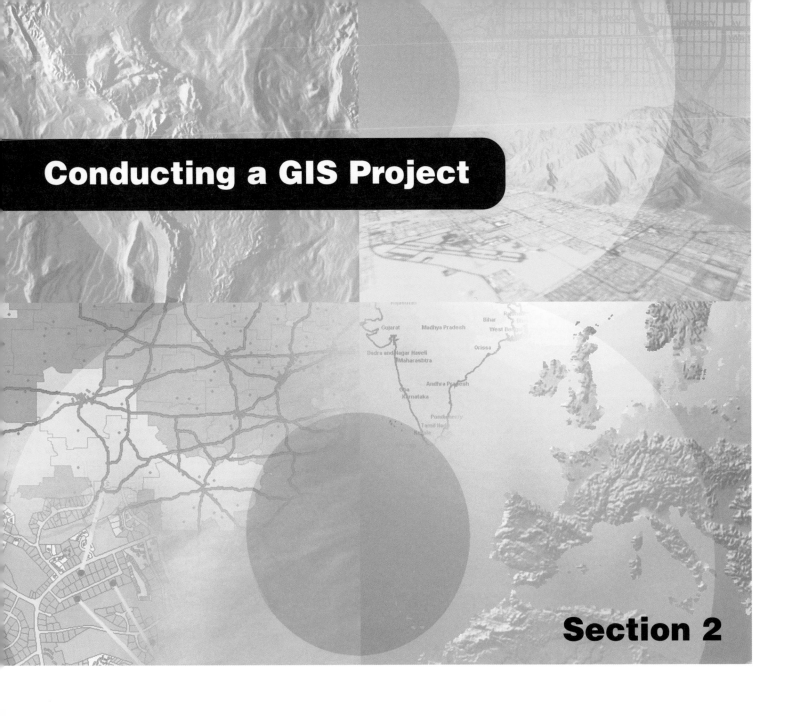

Conducting a GIS Project

Section 2

Planning a GIS project

4

Beginning with this chapter and through the rest of the book, you will conduct a sample GIS analysis project. The tasks you perform will help you learn the methods for performing your own GIS projects. You'll learn about several specific GIS analysis techniques and, perhaps more importantly, you'll learn how to plan and carry out a GIS project.

This chapter provides an overview of GIS analysis and presents the steps involved in conducting a GIS project. It then walks you through the first step—planning the project.

The scenario for the project involves finding the best site for a new wastewater treatment plant for the fictitious City of Greenvalley. To find a suitable site you will need to know the site selection criteria. You'll then need to identify the data needed to address these criteria and use the data to find suitable sites for the plant. These are fundamental elements of a GIS analysis project.

What is GIS analysis?

The phrase "GIS analysis" encompasses a wide variety of operations that you can do with a geographic information system. These range from simple display of features to complex, multistep analytical models.

Showing the geographic distribution of data

Perhaps the simplest form of GIS analysis is presenting the geographic distribution of data. This is conceptually the same as sticking pins in a wall map, a simple but powerful method of detecting patterns.

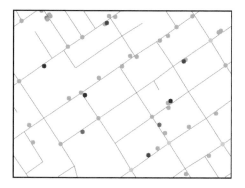

Here, the map is the analysis. A police department might analyze burglary patterns by plotting the addresses of reported break-ins. The department could make the map more informative by displaying the incidents with different symbols to show the time of day, method of entry, or types of valuables reported stolen.

Querying GIS data

Another type of GIS analysis is querying, or selecting from, the database. Queries let you identify and focus on a specific set of features. There are two types of GIS queries, *attribute* and *location* queries.

Attribute queries, also called aspatial queries, find features based on their attributes. The police department mentioned above could use an attribute query of their database to obtain a table of crimes that fall into a particular category.

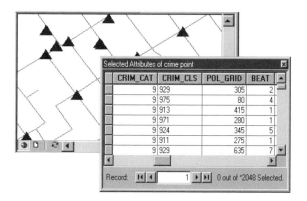

Here are the results from a query on the CRIM_CAT field, showing records where the value in the field is 9. The map shows the results of the query.

Location queries, also called spatial queries, find features based on where they are. The police department could use a location query of the database to find crimes that occurred within a given area.

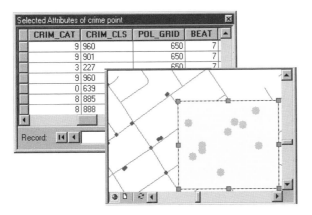

One way to do a location query is by drawing a rectangle on the map. Here, the police department has selected only those crimes that occurred within the rectangle. These crimes could be studied together to determine if any of them are related.

The police department could also do more complex spatial queries using polygon features, such as census tracts, selected from another layer. One of the most useful features of a GIS is that you can see the results of both spatial and aspatial queries on the map.

Identifying what is nearby

A third type of GIS analysis is finding what is near a feature. One way to find what is near a feature is by creating a buffer around the feature.

A city planning commission could identify the area within 1,000 meters of a proposed airport by buffering the airport feature. The buffer could be used with other layers of data to show which schools or hospitals would be near the new airport.

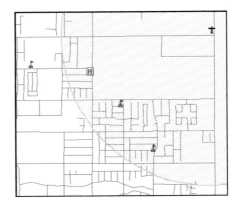

A powerful function of GIS analysis is that the output of one procedure can be used in another. Here, the buffer around the airport is used in a location query. Two schools and a hospital that are within the buffer have been selected. The school that is outside of the buffer was not selected.

Overlaying different layers

A fourth type of GIS analysis is overlaying different layers of features. You can create new information when you overlay one set of features with another. There are several types of overlay operations, but all involve joining two existing sets of features into a single new set of features.

For example, a farmer wants to find how much land can be planted with a new crop. The crop can't be planted on hillsides and needs soils that are highly permeable.

The farmer combines, in a *union* overlay, two existing layers of data about the farm: polygons of the ground surface classified by slope and polygons of soil permeability. The farmer can now select the new polygons that have low slopes and high permeability.

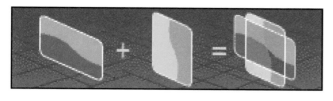

Slope   Permeability   Slope and
                                     permeability

There are several different spatial overlay and spatial manipulation operations that you can use on layers including union, intersect, merge, dissolve, and clip.

Doing a complex analysis

You can combine all of these techniques and many others in a complex GIS analysis. With a GIS you can create detailed models of the world to solve complicated problems. Because a GIS can perform these operations rapidly, it is possible to repeat an analysis using slightly different parameters each time and compare the results. This can allow you to refine your analysis techniques.

This section has provided a brief overview of some of the common types of GIS analysis. For more information on these and other kinds of analysis you can perform using GIS, see *The ESRI Guide to GIS Analysis*.

In the next section you'll learn about the steps in a typical GIS analysis project.

The steps in a GIS project

In a typical GIS analysis project, you identify the objectives of the project, create a project database containing the data you need to solve the problem, use GIS functions to create an analytical model to solve the problem, and present the results of the analysis.

Step 1: Identify your objectives

The first step of the process is to identify the objective of the analysis. You should consider the following questions when you are identifying your objectives:

- What is the problem to solve? How is it solved now? Are there alternate ways to solve it using a GIS?

- What are the final products of the project—reports, working maps, presentation-quality maps?

- Who is the intended audience of these products—the public, technicians, planners, officials?

- Will the data be used for other purposes? What are the requirements for these?

This step is important because the answers to these questions determine the scope of the project as well as how you implement the analysis.

Step 2: Create a project database

The second step is to create a project database. Creating the project database is a three-step process. The steps are designing the database, automating and gathering data for the database, and managing the database.

Designing the database includes identifying the spatial data you will need based on the requirements of the analysis, determining the required feature attributes, setting the study area boundary, and choosing the coordinate system to use.

Automating the data involves digitizing or converting data from other systems and formats into a usable format as well as verifying the data and correcting errors.

Managing the database involves verifying coordinate systems and joining adjacent layers.

Creating the project database is a critical and time-consuming part of the project. The completeness and accuracy of the data you use in your analysis determines the accuracy of the results.

Step 3: Analyze the data

The third step is to analyze the data. As you've seen, analyzing data in a GIS ranges from simple mapping to creating complex spatial models. A model is a representation of reality used to simulate a process, predict an outcome, or analyze a problem.

A spatial model involves applying one or more of three categories of GIS function to some spatial data. These functions are:

- Geometric modeling functions—calculating distances, generating buffers, and calculating areas and perimeters.
- Coincidence modeling functions—overlaying datasets to find places where values coincide.
- Adjacency modeling functions—allocating, pathfinding, and redistricting.

With a GIS you can quickly perform analyses that would be impossible or extremely time consuming if done by hand. You create alternative scenarios by changing your methods or parameters and running the analysis again.

Step 4: Present the results

The fourth step is to present the results of your analysis. Your final product should effectively communicate your findings to your audience. In most cases, the results of a GIS analysis can best be shown on a map.

Charts and reports of selected data are two other ways of presenting your results. You can print charts and reports separately, embed them in documents created by other applications, or place them on your map.

What's next?

Now that you have reviewed the steps in a GIS project, you are ready to begin planning your own project. The next section presents an overview of the steps for the Greenvalley wastewater treatment plant project. The first step—identifying the project objectives—is covered in this chapter. The rest of the steps comprise the remaining chapters in this book.

Planning your project

Planning is a critical step in any GIS project and can save you time and effort once you get to the database creation, analysis, and mapping steps. During the planning phase you identify the project objectives, define the criteria for the analysis, and identify the data required to support the analysis. You should also consider the approach you'll use for the analysis and what the final products of the project will be. Once you've done this, you can proceed to create the project database.

Throughout the rest of this book you will be working on a small GIS analysis project. In the process, you'll learn how to plan a GIS project and how to use ArcMap and ArcCatalog together to carry it out. While you'll be performing a specific type of analysis—finding a site for a new facility—the steps you'll follow in the project, and many of the specific tasks, will be applicable to a range of GIS projects. The scenario for the project is to find a suitable site for a new wastewater treatment plant.

The City of Greenvalley is growing. To support this growth, the City is building a new wastewater treatment and recycling plant. The City plans to use conservation and wastewater recycling to help meet its expected water needs.

The diagram to the right outlines the steps in a GIS project and shows where each step is covered in the remaining chapters in this book.

In this chapter, you'll carry out Step 1—identifying the project objectives. You'll also do some planning for the remaining steps.

Steps in a GIS project

Step 1: Identify the objectives—Chapter 4

Step 2: Create the project database

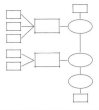

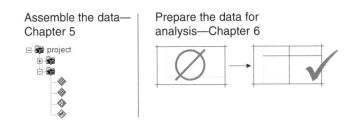

Step 3: Analyze the data—Chapter 7

Step 4: Present the results—Chapter 8

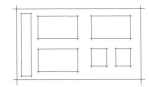

Step 1: Identify the project objectives

The objective of this GIS analysis is to find a suitable site for the City's new wastewater treatment plant. The City has never used a GIS model to site a wastewater treatment plant. The existing plant was sited many years ago using a quadrangle map, acetate overlays, and the City Council's knowledge of the area in consultation with the City engineer. This approach was adequate but time consuming, and the public was not involved in the process.

The problem has become more difficult as the area has become more developed and environmental and public health regulations more stringent. The Council has chosen to use a GIS model in order to speed the process and to ensure that the necessary regulations are complied with.

Because the Council recognizes that siting such a plant can be controversial, they want the analysis to identify all of the parcels that could be used for the plant site and then identify the highly suitable parcels, based on very specific criteria. The possible sites will be discussed at a public meeting. The map you create for the meeting should make clear which parcels are highly suitable, which are less suitable, and which are unsuitable.

The City has provided you with a list of the criteria for a suitable site. The parcels chosen for the site must be:

- Below 365 meters in elevation, to minimize pumping costs

- Outside of the floodplain, to avoid spillage during storms

- Within 1,000 meters of the river, to minimize pipeline construction for treated water that is discharged

- At least 150 meters from residential property and parks, to minimize the impact on the City's residents

- On vacant land that can be developed, to minimize land acquisition and construction costs

In addition, to further reduce construction costs, the City would prefer that the site be:

- Within 1,000 meters of the main wastewater junction (within 500 meters is considered even more suitable)

- Within 50 meters of an existing road

The plant will also require a total of at least 150,000 square meters in area.

A preliminary review of existing paper maps has shown that the most likely location for the plant is in the northwest corner of the City, near the river, and in a low-lying area. This will be the study area for the project. The GIS analysis will allow you to combine the criteria to identify specific parcels that are suitable sites.

Step 2: Create the project database

Creating the database for this project will be a two-step process. First you'll assemble the existing data and review it. Then you'll prepare the data for analysis. Some of the data will be usable as is; other layers will need additional processing. You may also need to automate some data. You'll assemble the data in Chapter 5, 'Assembling the database', and prepare the data for analysis in Chapter 6, 'Preparing data for analysis'. At this point, though, you can plan for those tasks by identifying the data layers you'll need and the sources of the data.

Assemble the project data

Several City of Greenvalley departments maintain GIS data and have working arrangements to share data for City projects. Some of this data is already stored in the City's

GreenvalleyDB database. The City also has data sharing agreements with several regional and state agencies.

Because a database containing much of the data you need already exists, you will not need to spend as much time on designing and automating your project database as you would otherwise. However, you will still need to do some database design work for your project database. You'll need to identify the dataset and any attributes required for each criteria. You'll then research the available data to see which layers will meet your needs.

Each of the City's criteria will require a layer of data for the analysis. Here is a list of the criteria and the corresponding datasets and attributes.

CRITERIA	DATASET	ATTRIBUTES
LESS THAN 365 METERS ELEVATION	ELEVATION	ELEVATION IN METERS
OUTSIDE THE FLOODPLAIN	FLOODPLAIN	N/A
WITHIN 1,000 METERS OF THE RIVER	RIVER	N/A
AT LEAST 150 METERS FROM RESIDENTIAL PROPERTY	PARCELS	LAND USE
AT LEAST 150 METERS FROM PARKS	PARKS	N/A
ON VACANT LAND	PARCELS	LAND USE
WITHIN 1,000 METERS OF THE WASTEWATER JUNCTION	WASTEWATER JUNCTION	N/A
WITHIN 50 METERS OF A ROAD	ROADS	N/A
AT LEAST 150,000 SQ. METERS	PARCELS	AREA IN SQUARE METERS

Note that the parcels dataset will be used for several criteria.

You can now take inventory of the data that you have and see which layers correspond to the required datasets. You can also identify other layers that you need to obtain or create.

To find areas below 365 meters elevation, you need a source of elevation data. A colleague at the State Department of Transportation (DOT) has provided an elevation grid. Because you simply need to know whether or not a parcel is below 365 meters, you will use a polygon of areas below 365 meters, which your colleague at DOT has created from the grid. This data is in a shapefile format.

To find parcels outside of the floodplain, you will use the City Planning Department's digital flood zone layer, stored as a feature class in the City's GreenvalleyDB geodatabase.

To identify areas within 1,000 meters of the river, you will first need a layer of the river. The County Water Resources Department has a shapefile of the river.

You will need a dataset of the parcels in your study area. The City Tax Assessor has a tiled database of parcels stored as shapefiles. Two of these tiles cover your study area. The parcel database includes a land use attribute that you will use to identify residential parcels (so you can buffer them to 150 meters) and vacant parcels. You will use the area attribute of the parcel shapefile to identify possible sites with an area of at least 150,000 square meters.

To find areas more than 150 meters from parks, you'll need a parks layer. The City Parks and Recreation Department has a feature class of existing parks, which is stored in the GreenvalleyDB geodatabase.

There is also a recently discovered historic site in the project study area. The City plans to develop a park around the site, but the proposed park boundary has not been placed into the park feature class yet. You will get this information into your project database by digitizing from a scanned image of the draft park boundary map.

To find parcels within 1,000 meters of the main wastewater junction, you will need a layer that includes the junction. The City Utility Department has a coverage of the wastewater mains and the junction.

To identify parcels that are within 50 meters of a road, you will use the existing streets feature class from the GreenvalleyDB geodatabase.

The table below lists the layers you'll assemble for the project database, based on the available data. The source and the format of each layer are also listed.

LAYER	SOURCE	FORMAT
ELEVATION	STATE DEPT. OF TRANSPORTATION	GRID
ELEVATION < 365 M	STATE DEPT. OF TRANSPORTATION	SHAPEFILE
FLOODPLAIN	CITY PLANNING DEPT.	GEODATABASE
RIVER	COUNTY WATER RESOURCES DEPT.	SHAPEFILE
PARCELS	CITY TAX ASSESSOR	SHAPEFILES (TILED)
PARKS	CITY PARKS AND RECREATION DEPT.	GEODATABASE
HISTORIC PARK	CITY PARKS AND RECREATION DEPT.	SCANNED IMAGE
WASTEWATER JUNCTION	CITY UTILITIES DEPT.	COVERAGE
STREETS	CITY STREETS DEPT.	GEODATABASE

The database will also include the scanned image of the historic park, which you'll use to digitize the new park. You'll also include the elevation grid, as you may want to display it on your final map.

In Chapter 5, 'Assembling the database', you'll assemble the data and organize it so it's easily accessible within a single project folder. You'll then review the data to see which layers will require additional processing.

Prepare data for analysis

Based on your review of the data, you'll determine which layers are currently usable and which require additional processing for use in the analysis. Some of the common tasks involved in preparing data for analysis include:

- Checking data quality (making sure the data is accurate and up-to-date)
- Converting data between formats
- Automating data by digitizing, scanning, converting, or geolocating
- Defining coordinate systems
- Projecting layers to a new coordinate system
- Merging adjacent layers

You'll need to perform some of these tasks for your project database. You already know, for instance, that the boundary of the proposed park surrounding the historic site will need to be digitized. You have a scanned map of the proposed boundary that you will register to the City's geodatabase and digitize using the parcels layer as a backdrop. The new park feature will be added to the existing park feature class in the GreenvalleyDB geodatabase.

You will also need to merge the two parcel tiles for your study area to make the analysis easier to perform.

Once you've reviewed the existing data (in Chapter 5), you'll be able to see which other layers require additional processing.

Most of the data for the project is already in coverage, shapefile, geodatabase, or raster format, all of which ArcGIS can use, simultaneously. There may be cases, though, where you'll need to convert data to a different format (for example, if converting from vector to raster format or from shapefile to geodatabase feature class, for storage in an existing geodatabase).

ArcGIS can display and overlay layers in different coordinate systems as long as the coordinate system for each layer is defined. You'll need to check this, especially for data you've obtained from other sources.

You'll perform the necessary data processing tasks in Chapter 6, 'Preparing data for analysis'.

Step 3: Analyze the data

During the planning stage of the project, you'll want to consider the analysis methodology and list the major steps in the process. That way you can make sure you are aware of all the datasets you will need and can include them as you create the project database. You may want to create a schematic diagram of the process as a guide.

The diagram to the right shows the process for the wastewater treatment plant siting analysis.

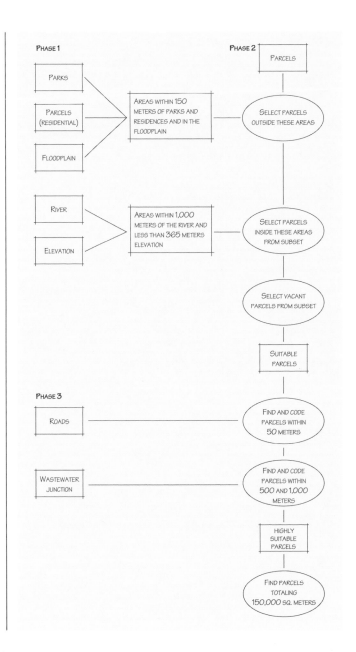

The analysis consists of three phases.

In the first phase, you'll create a layer of the areas the plant should be outside of and another layer of the areas the plant should be within.

In phase two, you'll use these layers to select a subset of parcels that are in a suitable location. You'll then select the subset of these that are vacant to create a layer of suitable parcels.

In the third phase, you'll consider the City's additional criteria that define the highly suitable parcels. You'll find the suitable parcels within 50 meters of a road and those within 500 and 1,000 meters of the wastewater junction, then tag them with the appropriate codes so they can be identified on the map. You'll also check to see which parcels are large enough for the construction of the plant.

While the schematic shows the major steps in the process, there are a number of interim steps you'll complete in each phase. You'll develop the detailed methodology and perform the analysis in Chapter 7, 'Performing the analysis'.

Step 4: Present the results

During project planning you should consider the purpose and audience for your final products. For this project, you'll present the results of the analysis on a presentation-quality map that shows the parcels that are suitable and highly suitable sites. The map will be presented to a general audience at a public meeting.

While you don't need to design the map layout at this point, you'll want to consider the layers that will be shown on the map. In addition to the analysis layers, you may want to include other layers that will provide context for the analysis results and make the map easier to read and understand.

For this project, in addition to the layers you'll use and create during the analysis, you'll want to show the elevation grid as a backdrop so map readers can see the areas of lower and higher elevation in the City, as elevation has a major impact on the location of the wastewater treatment plant.

You'll design and create the presentation map in Chapter 8, 'Presenting the results'.

What's next?

Now that you've reviewed the steps involved in a GIS project, identified the project objectives, and engaged in some project planning, it's time to get started.

You'll assemble the data for the project and review the data in the next chapter.

Assembling the database

5

IN THIS CHAPTER

- **Organizing the project database**
- **Adding data to the project folder**
- **Previewing the data in ArcCatalog**
- **Examining the data in ArcMap**
- **Cleaning up the Catalog tree**

The data you will need for the project exists in several places and in different formats. In order to conduct your analysis, you must find the data, get information about it, and copy it into the appropriate workspace. ArcCatalog lets you explore and organize your data efficiently.

In this chapter, you will organize your project database to hold the data that you will obtain or create. You will use ArcCatalog to preview data and copy it, create folders to hold data, and create layers to represent remote data. Organizing your project database in a single branch of the Catalog tree makes it easier to find the data you need, so you'll create a connection to the project folder.

You'll also use ArcMap to display the datasets in your project database, so you can see the geographic relationships between the various datasets you'll be working with during the analysis.

By previewing the data in ArcCatalog and ArcMap, you'll be able to see which layers will need additional processing to be usable for the analysis.

Organizing the project database

There are many ways to organize a project database. One good strategy is to create a single project folder, then subfolders to hold the input datasets, and another subfolder to hold the datasets you create during the analysis.

As with many GIS projects, the data for your project comes from several different sources. Some of it is in different data formats or in different coordinate systems. Most of the data has already been collected for you. Here is where the data currently resides:

Layer	Name	Format	Current Location
Elevation	ELEVATION	Grid	State_share folder
Elevation < 365 m	LOWLAND	Shapefile	State_share folder
Floodplain	FLOOD_POLYGON	Geodatabase	GreenValleyDB geodatabase
River	RIVER	Shapefile	County_share folder
Parcels	parcel_1, parcel_2	Shapefiles (tiled)	City_share\Land folder
Parks	PARKS_POLYGON	Geodatabase	GreenValleyDB geodatabase
Historic park	HISTORIC.TIF	Scanned image	City_share\Image folder
Wastewater junction	JUNCTION	Coverage	City_share\Utility folder
Streets	STREET_ARC	Geodatabase	GreenValleyDB geodatabase

You'll copy the data (to maintain the original as a backup) and organize it in a single project folder to make it more accessible. You'll also create a new folder to store the data you'll create during the analysis.

There is no single way to structure a project database; it partly depends on personal preference. The goal is to minimize duplication of datasets and to have the data well organized and easily accessible. This will help avoid confusion during the project as well as if you need to revisit the project in the future. Before you start to create the folders on disk and move the datasets around, it's a good idea to sketch out the organization of the project folders.

PROJECT FOLDER
— ANALYSIS FOLDER
— CITY_LAYERS FOLDER
— CITY_SHARE FOLDER
  — IMAGE FOLDER
  — LAND FOLDER
  — UTILITY FOLDER
— COUNTY_SHARE FOLDER
— STATE_SHARE FOLDER
— WATERPROJECT GEODATABASE

The City_share, County_share, and State_share folders are stored locally on your computer, but they could represent shared folders accessed over a network. You can use ArcGIS to manage and display GIS data on any shared drive on your network.

You'll also want to think about how to name the new datasets you'll create, and create naming conventions. Using meaningful names can help you see at a glance what each dataset is. For example, if you merge two parcel datasets, you might name the output parcel01mrg to indicate that it was the first parcel dataset created and the operation was a merge. If you then edit the dataset, you might name the edited version parcel02edt, and so on.

You'll use ArcCatalog to copy the folder containing the shared data to a new location so you can work with it while maintaining the original data. You'll then create the new personal geodatabase to hold several of the new datasets you'll create. You'll also create the two new folders: one to contain the layers from the City's GreenvalleyDB geodatabase and another to contain the layers you create during the analysis. Here are the steps:

- Copy the project folder.

- Create a connection to the project folder.

- Create the WaterProject personal geodatabase in the project folder.

- Create the City_layers folder in the project folder.

- Create the Analysis folder in the project folder.

If you have not already done the tutorial in Chapter 2, 'Exploring ArcCatalog and ArcMap', you will need to check with your system administrator to learn where the tutorial data is installed. Before you start the project you will also need to make a folder connection in ArcCatalog to the Greenvalley folder (use the instructions in Chapter 2).

Copy the project folder

The project folder contains data that other organizations are sharing with you. You'll copy the whole folder intact to your own drive. First, open ArcCatalog.

1. Click Start, point to Programs, point to ArcGIS, and click ArcCatalog.

2. Navigate to the ArcGIS\ArcTutor\Getting_Started folder. Double-click the Getting_Started folder to show its contents.

3. Click the project folder, hold the Ctrl key, drag the folder from its current location, and drop it onto your C:\ drive or any other local drive or folder.

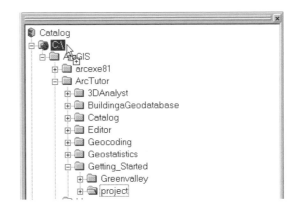

Substitute your local drive or folder for "C:\" for the rest of this chapter.

Dragging a folder to another drive (for example, from the C:\ drive to the D:\drive) copies the folder and its contents to the new location. Dragging to another location on the same drive moves the folder. To copy a folder to another location on the same drive, hold the Ctrl key while you drag.

4. When ArcGIS is finished copying the data, click C:\ in the Catalog tree to view the contents of the C:\ drive on the right side of the Catalog window.

You can see that the project folder is listed.

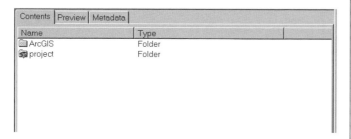

Now that you've copied the project folder, you can work on your copy of the data without modifying the original data.

Connect to the project folder

If you have many folders on a drive, it can become tedious to scroll to one you use frequently. Making a connection puts that folder at your fingertips. You'll create a connection for the project folder.

In the tutorial, you made a folder connection by clicking Connect to Folder and browsing. Here is a quicker way:

1. Navigate to the project folder in the right side of the Catalog window (the Contents tab should be selected).

2. Click the project folder and drag and drop it onto Catalog at the top of the Catalog tree.

The new folder connection—C:\project—is now listed in the Catalog tree.

The connection is a shortcut to the project folder. For the rest of the project, you'll access the data in the project folder using the connection.

Create a personal geodatabase

Next, you'll create a personal geodatabase within the project folder to store several of the updated and new datasets you'll create during the project. Using a geodatabase is an efficient way of storing, accessing, and managing data.

1. Click the project folder connection you just created to see its contents in the right side of the Catalog window.

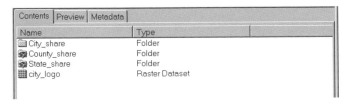

2. Right-click the project folder connection, point to New, and click Personal Geodatabase.

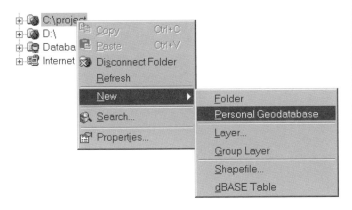

You will see additional options if you are using ArcInfo.

The new geodatabase is listed in the right side of the Catalog window with its name highlighted (New Personal Geodatabase).

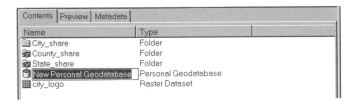

3. Rename the geodatabase by typing "WaterProject" over the highlighted text. Press Enter.

Create the City_layers and Analysis folders

Now you'll create two new folders under the project folder to hold the layers from the City's GreenvalleyDB geodatabase and the new layers you'll create later during the analysis.

1. Right-click the project folder, point to New, and click Folder.

The new folder is listed on the right side of the Catalog window with its name highlighted (New Folder).

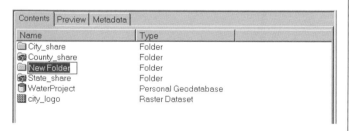

2. Rename the folder by typing "City_layers" over the highlighted text. Press Enter.

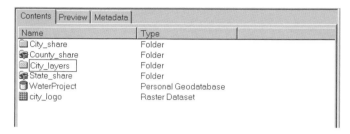

Create the Analysis folder the same way.

3. Right-click the project folder, point to New, and click Folder. Rename the folder "Analysis".

If you make a mistake and create the folder in the wrong location or misspell the name, simply right-click the folder, click Delete or Rename, and start over.

Adding data to the project folder

Three of the data sources you need—parks, streets, and the flood zone—are in the City's GreenvalleyDB database, which is already on your local drive. Since you'll be modifying the parks feature class by adding the new historic park, you'll copy it to the WaterProject geodatabase you just created. That way you maintain the original data as a backup. You won't be modifying the other two feature classes—only using them for display and analysis. Rather than copying them, you'll create layers in the project folder that point to the original data in the GreenvalleyDB geodatabase. That will let you access the data from within the project folder without creating duplicate datasets (this is especially useful when accessing data over a network). Here are the steps:

- Copy the parks feature class from the GreenvalleyDB geodatabase to the WaterProject geodatabase.
- Create the streets layer in the City_layers folder.
- Create the flood_zone layer in the City_layers folder.

Copy the parks feature class to the WaterProject geodatabase

1. Click the plus sign next to the project folder in the Catalog tree to expand the contents.

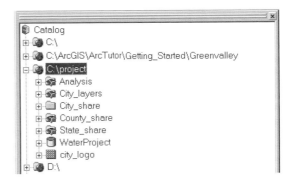

2. Double-click the Greenvalley folder connection in the Catalog tree.

   The contents appear in the right side of the Catalog window.

In the tutorial in Chapter 2, you created a connection to the Greenvalley folder. If the connection is no longer active, navigate to the Getting_Started folder, expand its branches, and then click the Greenvalley folder.

3. In the Catalog tree, double-click Data, double-click GreenvalleyDB, and click the Parks feature dataset.

The GreenvalleyDB geodatabase is organized using feature datasets such as hydrology and transportation. Feature datasets are useful for grouping related feature classes within a geodatabase. For example, you might include feature classes such as water mains, laterals, junctions, and pumps, within a feature dataset named WaterSystem. All the feature classes within a feature dataset have the same geographic extent. In addition, the feature classes maintain

some topological relationships in common. So, for example, if you edited the junction feature class and moved the location of a water junction, the attached lines in the mains and lateral feature classes would move accordingly.

Since the WaterProject geodatabase you created will contain only a few feature classes, it's not necessary to use feature datasets.

4. Click and drag the parks_polygon feature class to the WaterProject geodatabase in the Catalog tree (scroll down first if the WaterProject geodatabase isn't visible).

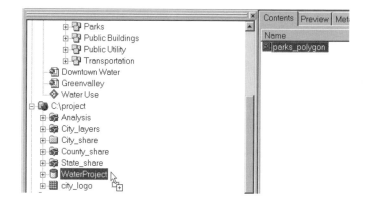

5. Click OK on the Data Transfer dialog box that appears.

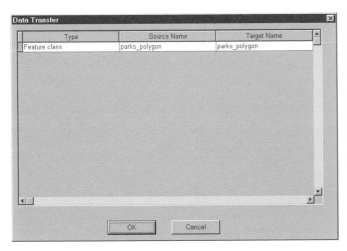

6. When the transfer is complete, click the plus sign next to the WaterProject geodatabase in the Catalog tree.

The parks_polygon feature class is listed.

Create the streets and flood zone layers

Unlike with the parks feature class, you won't be modifying the streets and flood zone data—you'll simply be using the data in the analysis process. So, rather than copying the data into the project folder, you'll create layers to serve as shortcuts to the data. That way you maintain a single copy of the data in the GreenvalleyDB but will be able to access the data from within the project folder.

The contents of the GreenvalleyDB geodatabase should still be visible in the Catalog tree. If not, double-click the Greenvalley folder to show its contents, then double-click Data, and double-click GreenvalleyDB.

1. Double-click Transportation.

2. Right-click street_arc and click Create Layer.

3. In the Save Layer As dialog box, navigate to the City_layers folder in your project folder and name the layer "streets". Click Save.

The streets layer is created in the City_layers folder.

Use the same procedure to create a layer for the flood zone data.

1. Double-click the Hydrology feature dataset in the Catalog tree, right-click the flood_polygon feature class, and click Create Layer.

2. Navigate to the City_layers folder in your project folder and name the layer "flood_zone". Click Save.

3. Click the project folder in the Catalog tree and double-click the City_layers folder.

The two layers are listed (you may need to click Refresh on the View menu to see them).

The streets and flood_zone layers are now stored with your project data. The actual data for each layer is stored in the GreenvalleyDB database. This database is on your local drive in this case, but it could just as easily be a remote database accessed over a network.

At this point, you've organized all the existing project data. Now you can access all the data from within the project folder. Your project folder should look like this:

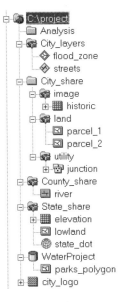

You may need to open each folder and the geodatabase to see all the layers.

Previewing the data in ArcCatalog

So far you've organized the project data by copying data folders and files. It's a good a idea to review each dataset to see what the spatial data looks like and what attributes it has. Doing so will help you make sure you've assembled the datasets you need. You will also be able to determine if any of the data needs additional processing to be usable for the analysis.

You have several options for examining the data. ArcCatalog lets you quickly preview the features and attributes in each individual dataset. ArcMap lets you display the datasets together, change how they're displayed, and zoom in on them to take a closer look. You'll use both ArcCatalog and ArcMap to review your data.

Preview the streets and flood zone layers

1. Navigate to and click the flood_zone layer in the Catalog tree.

The right side of the Catalog window displays the name of the layer, along with its type and a gray square

containing an icon representing the flood_zone polygons.

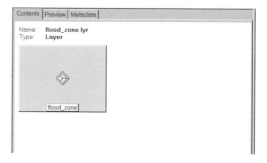

2. Click the Preview tab.

The flood_zone polygons are displayed.

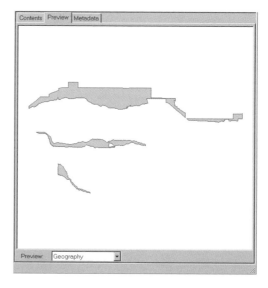

While you're here, you can create a thumbnail view of the flood_zone layer. That will let you quickly see what the layer looks like as you navigate through the Catalog tree. Since you just created the layers, the thumbnails don't exist yet.

3. Click the Create Thumbnail button on the toolbar.

Nothing happens on the screen, but the thumbnail is created and stored with the flood_zone layer.

—Create Thumbnail

4. Click the Contents tab.

Rather than the gray square you saw previously, you see the thumbnail showing the actual flood_zone polygons.

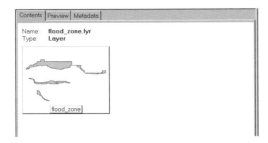

Now preview and create a thumbnail for the streets layer.

5. Click the streets layer in the Catalog tree and click the Preview tab.

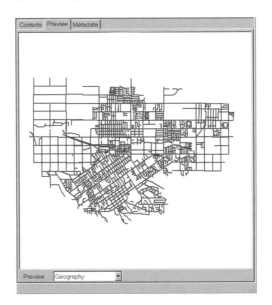

The streets are displayed.

6. Click the Create Thumbnail button on the toolbar.

7. Click the Contents tab to see the thumbnail.

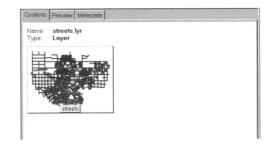

8. Click the City_layers folder in the Catalog tree, then click the Thumbnails button on the toolbar.

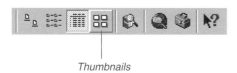

Thumbnails

The thumbnails you just created for these layers are displayed.

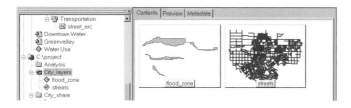

Thumbnails are useful for quickly previewing all the spatial data in a particular folder or geodatabase.

Explore the other data folders

You can preview the other datasets by looking at the contents of the other folders in the project database. The thumbnails for these datasets already exist.

1. Click the plus sign next to the City_share folder in the Catalog tree to list its contents.

2. Click the image folder.

   Since the Contents tab is selected and the Thumbnail button on the toolbar is also selected, the thumbnail for the historic park TIFF file displays.

3. Click the land folder to see the two parcel shapefiles you'll be working with, then click the utility folder to see the water junction coverage.

4. Click the State_share folder to display the thumbnails for the elevation grid and lowland shapefile.

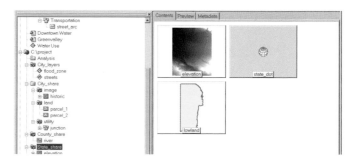

The State_share folder also contains a file named state_dot.prj that contains the coordinate system definition for Department of Transportation data. Your colleague at the State DOT wisely included it in case there was a question about the coordinate system of the elevation datasets. Since it's not a geographic dataset, there is no thumbnail.

Preview the river shapefile

The County_share folder contains the river shapefile created by the County Water Resources Department.

1. Double-click County_share to show the folder's contents, if necessary.

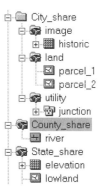

The river thumbnail appears. The shapefile contains a single river. You can take a closer look to make sure they sent you the right shapefile.

2. Click the Preview tab.

You get a message saying "The selection cannot be previewed." That's because the County_share folder is still selected.

The Preview tab previews a single dataset only. In contrast, the Contents tab displays all the datasets in a folder or geodatabase (either as thumbnails, as a list, or as icons representing the datasets). The Contents tab will also show you the contents of a single dataset. For shapefiles, feature classes, or image files, it displays the name and type of the dataset, along with the thumbnail. For coverages, the Contents tab lists the files comprising the coverage.

3. Click the river shapefile in the Catalog tree.

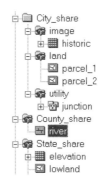

Now the river is displayed. You can check its attributes to be sure this is the right river.

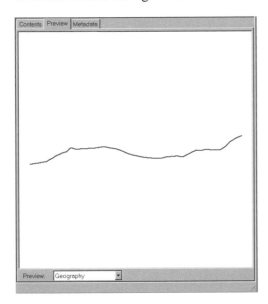

4. Click the Preview dropdown arrow and click Table.

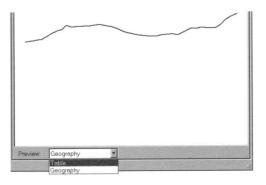

Now you can see the feature attribute table for the river shapefile. Your project area is along the Green River. This is the right shapefile.

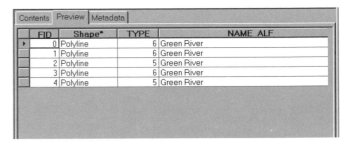

You're finished previewing the data in ArcCatalog for now.

5. Click the Contents tab.

Examining the data in ArcMap

The Contents and Preview tabs in ArcCatalog give you a quick overview of what your data looks like. However, the thumbnail images aren't drawn to scale or in the correct geographic space. Plus, each dataset is displayed separately. To see how the datasets relate to each other geographically, you need to display them in ArcMap. This will let you confirm that the datasets all overlap with your study area.

Open a new map

1. Click the Launch ArcMap button in ArcCatalog to start ArcMap.

Launch ArcMap

2. If the startup dialog box appears, click the option to use A new empty map and click OK.

If the startup dialog box doesn't appear, ArcMap automatically opens a new map.

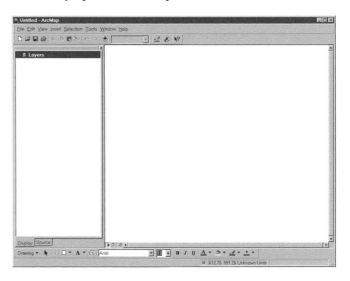

You can think of ArcMap as providing two functions: a workbench, or desktop, where you view, query, edit, and combine geographic data; and a canvas, or page, where you compose (layout) maps for display. You usually work in data view for the former and layout view for the latter (although you can also work with the data in layout view). Data view is the default when you start a new map. You'll do most of the work for your project in data view. You'll work with layout view in Chapter 8, 'Presenting the results'.

Add the parcel layers to the map

The two parcel layers you've obtained from the City assessor's office constitute the area in which you'll focus your search for a wastewater treatment plant site. You'll combine the two layers in the next chapter, but for now you can display them together to get a better sense of the area you'll be working in.

There are two ways to add data to a map. You can use the Add Data button on the ArcMap toolbar and navigate to the location of the dataset, or you can drag datasets from ArcCatalog and drop them on the map. The end result is the same, so which you use is a matter of preference. You'll get a chance to use both in this section.

1. Click the Add data button on the ArcMap toolbar.

Add Data

2. Navigate to the City_share folder under the project folder.

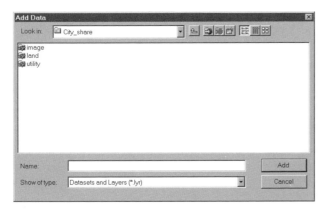

3. Double-click the land folder.

4. Click the parcel_1 shapefile, then press the Shift key and click the parcel_2 shapefile so both are selected.

5. Click Add.

The parcels are listed in the table of contents and displayed on the map. You can see that they're adjacent.

When you add a dataset to a map, ArcMap uses a color that it selects. The colors on your map may not match the ones shown here. You can change the colors and symbols for the layers on your map, as you'll see later in this chapter.

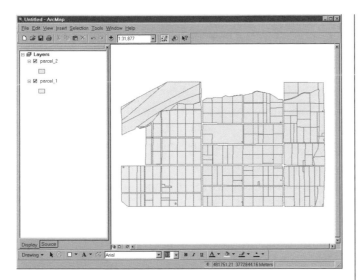

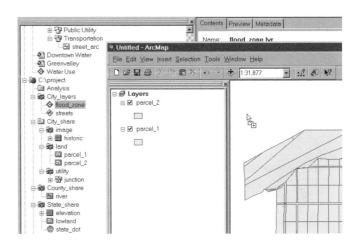

The flood_zone feature class is drawn on the map.

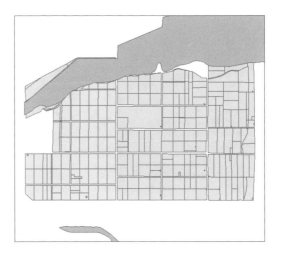

Add the rest of the City data to the map

Now you'll add to the map the streets and flood zone layers from the City_layers folder, the parks feature class from the WaterProject geodatabase, and the wastewater junction coverage from the City_share folder. As an alternative to adding data to a map using the Add data button, you can add data simply by dragging the datasets from ArcCatalog.

1. Make sure both the ArcCatalog and ArcMap windows are visible.

2. In the Catalog tree, navigate to the project folder.

3. Double-click City_layers, if necessary, to display its contents.

4. Click flood_zone and drag it onto the map.

5. Click the streets layer and drag it onto the map.

6. In the Catalog tree, open the utility folder under the City_share folder. Click and drag the junction coverage onto the map.

7. Finally, open the WaterProject geodatabase, if necessary, and click and drag the parks_polygon feature class onto the map.

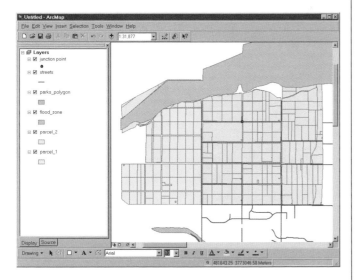

Now you've displayed most of the project datasets, stored in various folders and in different formats, all on the same map.

By default, ArcMap draws point features (such as the junction) on top of line features (such as the streets). It draws polygon features under the others. Within each type, the drawing order depends on the order the data was added to the map, with the most recent addition on top. You can rearrange the drawing order by clicking and dragging layers up or down in the table of contents. You can also choose your own colors and symbols for the layers.

Since the flood_zone layer is drawn on top of the parcels, it partially obscures them. You'll display the parcel outlines on top of the flood_zone so you can see the flood_zone beneath.

8. In the ArcMap table of contents, click the flood_zone layer and drag it to the bottom.

9. Right-click the legend symbol under parcel_1.

10. Click No Color at the top of the color palette.

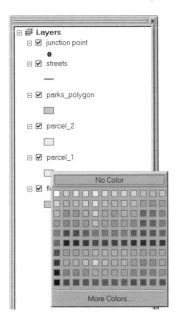

Do the same for the parcel_2 coverage.

11. Right-click the legend symbol under parcel_2 and click No Color on the color palette.

The parcel outlines are displayed, and you can see the flood zone area underneath.

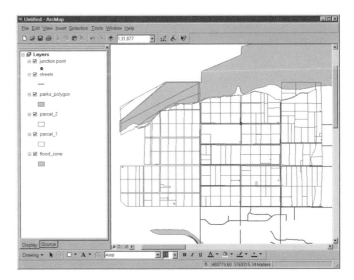

You're still zoomed in to the study area delineated by the parcel coverages. To get the big picture, zoom to the full extent of the datasets.

12. Click the Full Extent button on the Tools toolbar.

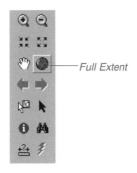

Now the map extent includes the additional datasets you've added to the map. You can see where the parcels you're focusing on are in relation to the rest of the City.

Add the river shapefile to the map

Next you'll add the river shapefile, from the County Water Resources Department, to the map.

1. Click the Add Data button on the ArcMap standard toolbar.

2. Navigate to the County_share folder under the project folder.

3. Click river.shp and click Add.

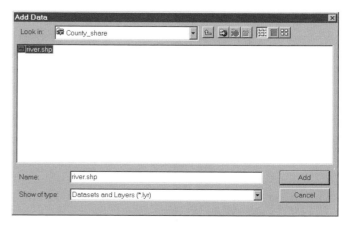

ArcMap displays a message that the river shapefile has a geographic coordinate system that differs from other data in the map. All the datasets you've added up to this point are from the City, and all use the same geographic coordinate system (Transverse Mercator). Apparently, the river shapefile uses a different coordinate system than the City data.

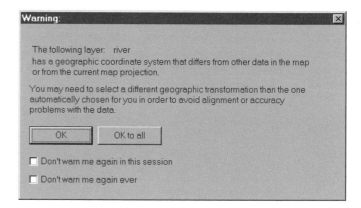

Every dataset uses a geographic coordinate system to link the coordinates stored in the GIS to actual locations on the surface of the earth. The coordinate system allows the GIS to determine where geographic features are located in relation to each other. There are many different coordinate systems used with geographic data. You can display the coordinate system for a dataset in ArcCatalog. ArcToolbox lets you define or change the coordinate system for a dataset. You'll get a chance to do this in the next chapter, where you'll also learn more about what a coordinate system is.

When you create a new map in ArcMap, the coordinate system used by the first dataset you add to the map, in this case the parcel_1 shapefile, defines the coordinate system for the entire map. If you add a dataset that doesn't use that same coordinate system, ArcMap will attempt to transform it on the fly so it displays correctly with the other data.

4. Click OK to close the Warning message box.

ArcMap transforms the river to the same coordinate system as the other datasets so it can be displayed.

You could just leave the shapefile in its original coordinate system, but since it will eventually be added to the City's GreenvalleyDB geodatabase, you'll want to put it in the same coordinate system as the rest of the City's data. You'll do that in the next chapter.

Add the elevation data to the map

Next you'll take a look at the elevation data from the State Department of Transportation.

1. Click the Add data button and navigate to the State_share folder.

2. Click elevation and click Add.

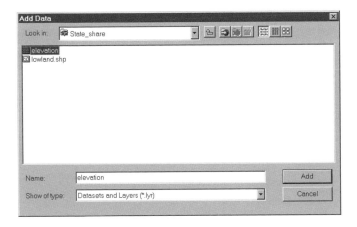

As with the river shapefile, ArcMap warns you that this dataset has a coordinate system that differs from other data on the map. Apparently the elevation data is also in a coordinate system different than that used by the City for its data.

3. Click OK to close the Warning message box. ArcMap adds the elevation grid to the map.

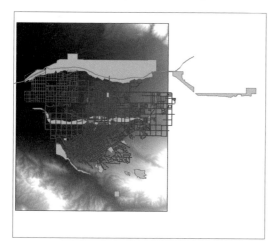

Since the grid is raster data, which is displayed as a continuous layer, it is added to the bottom of the table of contents and is drawn beneath the other layers. You can see that the grid doesn't underlie the entire City but does coincide with your study area.

Now add the lowland shapefile to the map.

4. Click the Add data button and click lowland.shp, then click Add.

ArcMap warns you that the layer is missing spatial reference information and can't be projected.

ArcMap can only transform data on the fly if the data is in latitude–longitude (which ArcMap can recognize) or if the coordinate system for the data has been previously defined, as it was for the elevation grid.

Apparently, the coordinate system was not defined for the lowland shapefile when it was created from the elevation grid. Presumably, it is in the same coordinate system as the elevation grid, but ArcMap doesn't know that at this point.

5. Click OK to close the Warning box.

ArcMap adds the data. You'll notice that while lowland appears in the table of contents, it doesn't appear on the map (it should appear on top of the layers that are listed below it in the table of contents). That's because it's in another—unknown at this point—coordinate system, so it can't be displayed correctly with the other data.

6. Click the Full Extent tool.

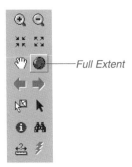

—Full Extent

There is a tiny dot at the bottom of the screen (that's the lowland shapefile) and a slightly larger dot at the top of the screen—that's the rest of the data. ArcMap adjusts the extent to the full range of coordinate values in both coordinate systems and draws all the layers within that extent.

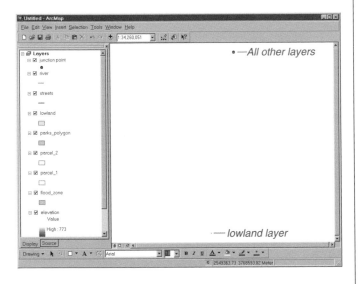

7. Right-click lowland in the table of contents and click Zoom To Layer.

Now you can see the lowland shapefile but not the rest of the data. You'll define the coordinate system for lowland in the next chapter so you can display and overlay it with the other data.

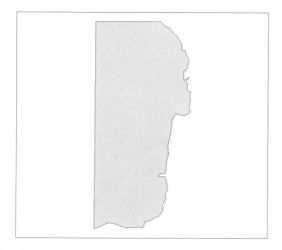

8. Click the Go Back to Previous Extent tool, then click it again.

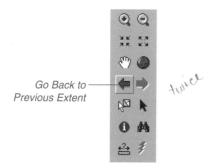

Go Back to Previous Extent

twice

The other datasets should now be displayed.

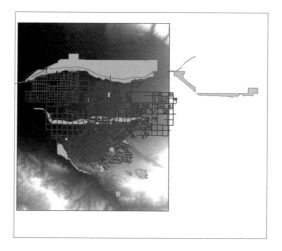

Create a layer from the elevation grid

By default, the elevation grid is displayed using a grayscale. You'll be displaying the elevation grid on the final map later in the project, so while you're here, you'll create a new layer with the symbology you want to use.

1. Right-click elevation in the table of contents and click Properties.

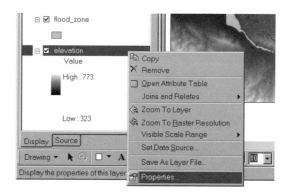

The Properties dialog box appears.

2. Click the Symbology tab.

3. Click the Color Ramp dropdown arrow, scroll down to the color ramp appropriate for representing elevation (from orange through yellow and green to blue), and click it.

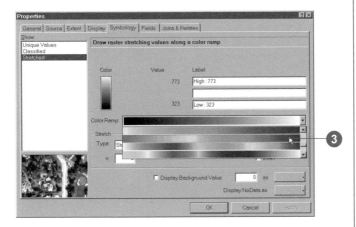

The default for this ramp is orange for low values and blue for high. You want to reverse the color ramp for the elevation grid.

4. Check the Invert check box.

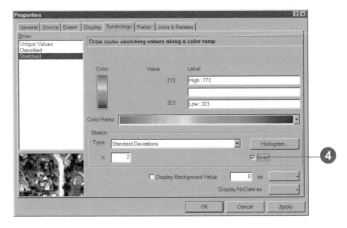

5. Click the Display tab and type 50 in the Transparent text box.

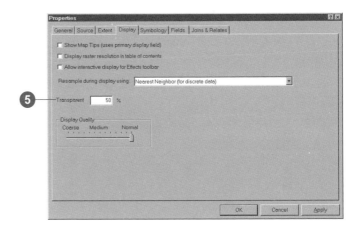

This will make the colors less intense, so it will be easier to see the other layers displayed on top of the grid.

6. Click OK. The grid is displayed using the color ramp and transparency settings you specified.

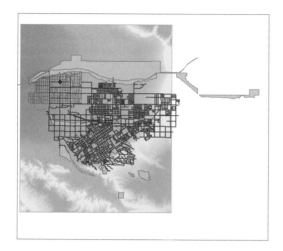

So far, the display settings for the elevation grid are valid only for the current map. To make sure the grid is displayed this way when you create your final map, you'll save it as a layer file.

7. Right-click elevation and click Save As Layer File.

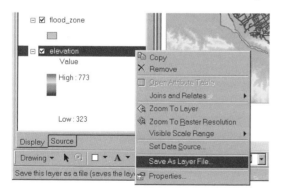

The Save Layer dialog box appears.

8. Navigate to the City_layers folder, name the layer elevation_grid, and click Save.

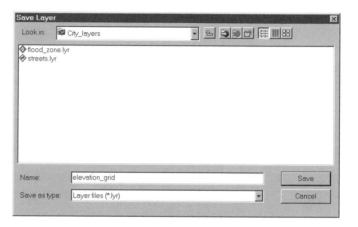

Now when you want to display the elevation grid again, you can just add the elevation_grid layer to the map and it will be drawn using the display settings you specified.

Layers store a shortcut to a data source and, optionally, information about how the data should be displayed on the map. Any time you add a dataset to a map in ArcMap, you create a layer since the map stores the data source and the symbology associated with it. When you save the map, the layer information is also saved.

As you've seen, you can also create layers as separate files. These layer files can be used to store symbology and other information, so the data is displayed the same way each time it is added to a map (as you just did with the elevation grid layer). They can also be used to access a data source without having to navigate to the actual location of the data (as you did earlier with the streets and flood_zone layers).

Save your map

The map you've been using to assemble your project database is a working map. You'll be displaying and using some of these same layers in the next chapter. Save the map now so you can use it in the next chapter and won't need to add the layers again.

1. Click File and click Save.

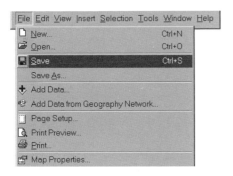

2. Navigate to the project folder.

3. Type "wastewater data" in the File name text box.

4. Click Save.

The map is saved as a map file. Notice that the name of the map now appears in the title bar.

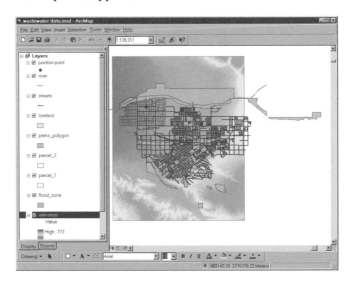

You're done using ArcMap for the time being, so go ahead and close it.

5. Click File and click Exit, or simply click the Close button (x) in the upper-right corner of the ArcMap window.

Cleaning up the Catalog tree

At this point, you've made folder connections, created and copied folders, and created layers to organize your project database. The Catalog tree in ArcCatalog is beginning to appear cluttered. Before you begin working with the data in the next chapter, clean up the Catalog tree. This will make it easier to find the data you need later on.

The connection to the tutorial folder you created in Chapter 2 is unnecessary now that you have copied the project folder and the files supplied by the City, County, and State. Deleting this connection will make more room in the tree.

1. Right-click the ArcGIS\ArcTutor\Getting_Started\ Greenvalley folder connection and click Disconnect Folder.

The folder connection is removed from the Catalog tree.

Now the Catalog shows only the data that you need for the project. (You may need to click View and click Refresh or close and reopen ArcCatalog to see the elevation_grid layer and the wastewater data map you created since you last used ArcCatalog.)

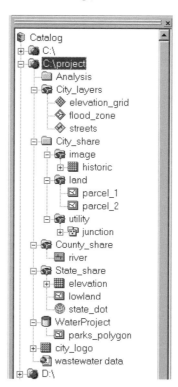

In this chapter, you have assembled the available data into an organized project database and reviewed the data. Some of the data will need additional processing before you can use it in the analysis. The two adjacent parcel tiles will need to be merged into one layer. You'll also need to transform the river shapefile into the same coordinate system as the rest of the City's data since it will become part of the City's permanent database. Since the elevation data will only be used for this project, you don't need to transform the data, but you do need to define the coordinate system for the lowland shapefile. Finally, you'll recall from the project planning in Chapter 4, 'Planning a GIS project', that you'll need to update the parks layer to include the new historic park. Here's a list of the layers you'll be working with, their new location in the project folder, and the processing required for each.

You'll undertake the data preparation tasks in the next chapter. If you're continuing on, keep ArcCatalog open.

Name	Format	Location	Processing
ELEVATION	GRID	STATE_SHARE folder	NONE
LOWLAND	SHAPEFILE	STATE_SHARE folder	DEFINE COORDINATE SYSTEM
FLOOD_ZONE	LAYER FILE (FROM GEODATABASE)	CITY_LAYERS folder	NONE
RIVER	SHAPEFILE	COUNTY_SHARE folder	DEFINE COORDINATE SYSTEM; PROJECT TO CITY'S COORDINATE SYSTEM; EXPORT TO GEODATABASE
PARCEL_1, PARCEL_2	SHAPEFILES (TILED)	CITY_SHARE\LAND folder	MERGE LAYERS
PARKS_POLYGON	GEODATABASE	WATERPROJECT GEODATABASE	UPDATE WITH NEW HISTORIC PARK
HISTORIC.TIF	SCANNED IMAGE	CITY_SHARE\IMAGE folder	DIGITIZE INTO PARKS FEATURE CLASS
JUNCTION	COVERAGE	CITY_SHARE\UTILITY folder	NONE
STREETS	LAYER FILE (FROM GEODATABASE)	CITY_LAYERS folder	NONE

Preparing data for analysis

6

IN THIS CHAPTER

- **Defining a coordinate system**
- **Projecting a shapefile**
- **Exporting a shapefile to a geodatabase**
- **Digitizing a new feature**
- **Merging two datasets**

Now that you have collected and organized the available data, you'll need to prepare the data for analysis. Some of your data is usable as is, but some needs additional processing. Making data usable for analysis can encompass a variety of tasks.

Unless GIS data is in the same coordinate system, it will not display or overlay correctly. ArcMap is able to match the coordinate systems of two different data sources, so you can display them together, as long as the coordinate systems are defined for both. However, if the data will become part of a permanent GIS database, you'll want to make sure it is in the same coordinate system and same data format as the rest of the database.

You may need to update or modify existing features based on more recent information. This can include modifying or adding spatial features or changing or adding values in a dataset's attribute table.

Features are sometimes stored as sets of adjacent tiles such as map sheets in a series. For analysis, it's often easier to join adjacent datasets into a single set of features so you can work with all the features at once.

You may also need to obtain new data for your project based on the requirements of the analysis. In some cases you may be able to get data in a usable format from another organization locally or on the Internet (either for free or by buying it from a commercial source). In other cases, you'll need to create the data by digitizing or scanning from a paper map or converting the data from a table or list (such as a list of customer addresses).

Data preparation tasks

For this project, you'll need to perform several tasks to prepare your data for analysis. You'll be working with data from various sources and in various formats: shapefiles, geodatabase feature classes, coverages, and rasters. ArcGIS lets you display and combine data in these formats without converting it. You'll define the coordinate system for the lowland shapefile so it can be displayed and combined with the other data. Then you'll project the river shapefile to the same coordinate system as the City's existing data and export it to the WaterProject geodatabase so it's ready to be placed into the City's geodatabase. The section "What are coordinate systems?" later in this chapter provides a brief overview of coordinate systems and map projections.

You'll also update the parks feature class with the new historic park so it, too, is ready to be placed back into the City's geodatabase. Finally, you'll merge the two parcel layers that comprise your study area.

Here are the processing steps to prepare the data for analysis:

- Define the coordinate system for the elevation data.
- Project the river shapefile to the City's coordinate system.
- Export the river shapefile to the WaterProject geodatabase.
- Digitize the historic park into the parks feature class.
- Merge the parcel layers.

You'll primarily be working with shapefiles since that's how most of the data came to you, but you'll also be working with data in the WaterProject personal geodatabase. A personal geodatabase is useful for processing data on a local computer that will become part of a large, multiuser geodatabase.

Defining the coordinate system for the elevation data

The elevation grid and lowland file are in a different coordinate system than the rest of the data. That's not a problem as long as the coordinate system is defined for those datasets. Without this information, however, ArcMap can't do a geographic transformation and the data can't be displayed or overlaid with the other project data. While the coordinate system for the elevation grid is defined, when the lowland shapefile was created from the grid, the coordinate system information was not included. You need to define the coordinate system for the shapefile.

If you closed ArcCatalog at the end of Chapter 5, 'Assembling the database', you'll need to reopen it now.

Check the coordinate system information

Before defining the coordinate system for the lowland shapefile, you'll check the coordinate system definitions for the City data and for the elevation grid. You'll do this by examining the metadata for the datasets.

1. In ArcCatalog, navigate to the WaterProject geodatabase under the project folder in the Catalog tree.

2. Open the database to list its contents, then click the parks_polygon feature class.

This feature class, which you copied from the City's GreenvalleyDB, is in the same coordinate system as the rest of the City's data.

3. Click the Metadata tab.

4. Click the Spatial tab in the metadata panel.

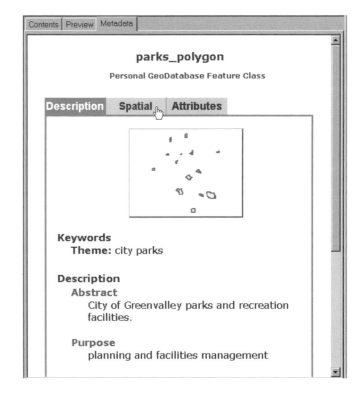

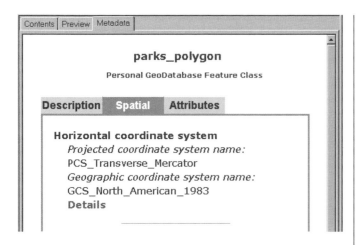

You can see that the coordinate system for the parks_polygon feature class uses a Transverse_Mercator projection.

Metadata contains information about each dataset. Some of the information is assigned and maintained automatically by ArcGIS; other information you add interactively. Metadata is invaluable when sharing datasets and for documenting GIS projects.

In this project you'll use metadata to get information you need for several of the steps. Metadata lets you store a great deal of information about a dataset: source, processing status, data quality, attribute values, and much more. For several of the datasets, we've provided some of the key information you'll need for the project.

In an actual GIS project you'd also use metadata to keep track of modifications you make to existing datasets and to document new datasets you create during the project. Adding or updating the metadata for a dataset takes a little extra time, but it pays off if you need to reuse the dataset in the future, share it with another department or organization, or reconstruct your processing steps.

Now check the coordinate system for the elevation grid.

5. Navigate to the State_share folder in the Catalog tree, open it, and click elevation.

6. Click the Spatial tab (when you select a new dataset, ArcCatalog defaults the metadata display to the Description tab).

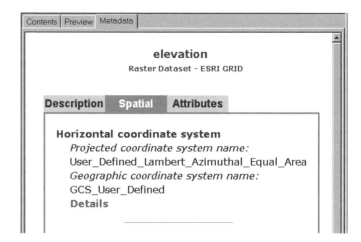

You can see that the elevation grid is in a coordinate system that uses a Lambert_Azimuthal_Equal_Area projection.

Finally, check the coordinate system information for the lowland shapefile.

7. In the State_share folder, click lowland.

8. Click the Spatial tab.

The metadata lists the bounding coordinates for the shapefile but doesn't list the coordinate system since it is unknown.

When you previewed the data in Chapter 5, 'Assembling the database', ArcMap was able to transform the elevation grid on the fly and display it with the other project data since the coordinate system was defined and stored with the grid. Because the coordinate system for the lowland shapefile is unknown, ArcMap was unable to transform it.

Define the coordinate system for the lowland shapefile

Presumably, the coordinate system for the lowland shapefile is the same as for the elevation grid since the shapefile was originally created from that grid. But you don't know for sure. Thoughtfully, the analyst at the Department of Transportation who sent you the data also included a spatial reference file that defines the coordinate system the department uses for all its data. You'll use the file state_dot.prj to define the coordinate system for the shapefile within ArcCatalog.

1. Right-click lowland in the Catalog tree and click Properties.

The Shapefile Properties dialog box appears.

2. Click the Fields tab.

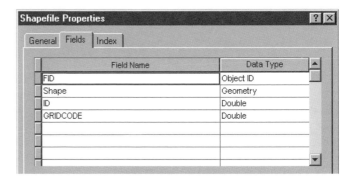

The fields in the shapefile's attribute table are listed. The Shape field contains the coordinate information for the shapefile.

3. In the Field Name list, click the row containing Shape.

The properties for the Shape field are displayed below in the Field Properties list. You can see that the Spatial Reference property is listed as Unknown.

4. Click the button with the ellipses (…) to the right of Spatial Reference.

The Spatial Reference Properties dialog box appears.

You'll be defining the coordinate system by selecting a predefined coordinate system—specifically, the one contained in the state_dot.prj file that accompanied the elevation and lowland data.

5. Click Select.

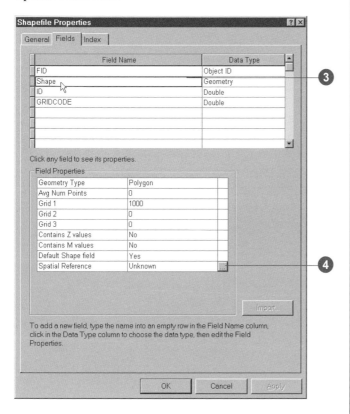

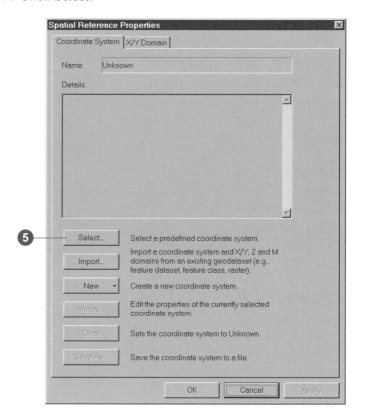

6. Navigate to the State_share folder under the project folder connection, click state_dot.prj, and click Add.

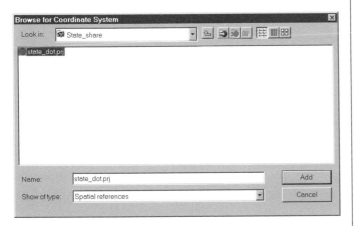

The name of the coordinate system appears in the Spatial Reference Properties dialog box, and the details are listed. You can see that they're the same as for the elevation grid.

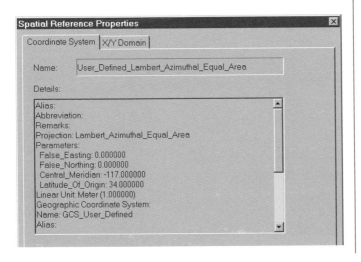

7. Click OK to close the Spatial Reference Properties dialog box.

The name of the coordinate system now appears in the Field Properties list.

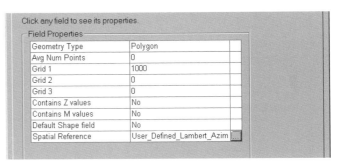

8. Click OK to close the Shapefile Properties dialog box.

You can verify the new coordinate system in the metadata.

9. Click View and click Refresh, then click the Spatial tab.

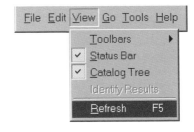

You can see that the coordinate system for the lowland shapefile is now defined.

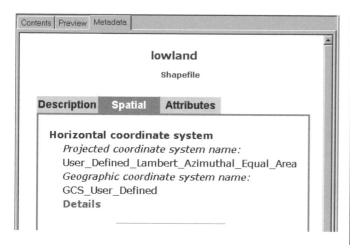

With the coordinate system defined, the lowland shapefile should now display correctly with the other project data and can be used in overlay operations during the analysis. You can check this in ArcMap.

10. Click the Launch ArcMap button on the toolbar.

Launch ArcMap

11. In the startup dialog box, click "wastewater data.mxd" and click OK (if the startup dialog box doesn't appear, click File on the ArcMap toolbar and click "wastewater data.mxd").

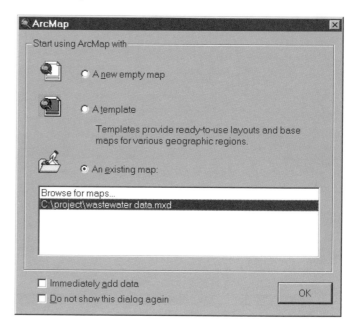

The lowland layer now appears in the same geographic space as the other project data.

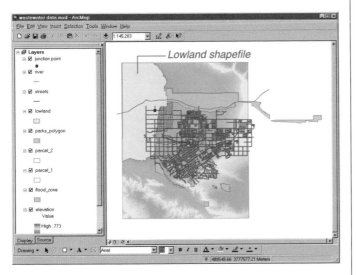
Lowland shapefile

12. Click lowland in the table of contents and drag it to the bottom so it displays beneath the elevation grid.

You can see the layer under the elevation grid and see that it does in fact encompass the lowest elevations in the City.

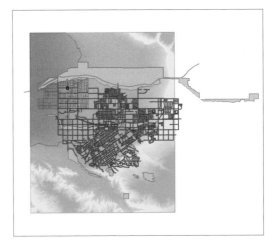

13. Close ArcMap. You won't be using this map again, so click No when prompted to save changes to the map.

What are coordinate systems?

ArcInfo stores features using x,y coordinates. These coordinates are linked to real-world locations by a coordinate system. The coordinate system specifies a datum and a map projection.

Datum

A *datum* is a mathematical representation of the shape of the earth's surface. A datum is defined by a spheroid, which approximates the shape of the earth and the spheroid's position relative to the center of the earth. There are many spheroids that represent the shape of the earth and many more datums based on them.

A horizontal datum provides a frame of reference for measuring locations on the surface of the earth. It defines the origin and orientation of latitude and longitude lines. A local datum aligns its spheroid to closely fit the earth's surface in a particular area; its origin point is located on the surface of the earth. The coordinates of the origin point are fixed, and all other points are calculated from this control point. The coordinate system origin of a local datum is not at the center of the earth. NAD27 and the European Datum of 1950 are local datums.

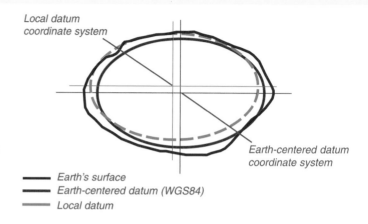

Local datum coordinate system

Earth-centered datum coordinate system

—— Earth's surface
—— Earth-centered datum (WGS84)
—— Local datum

In the last 15 years, satellite data has provided geodesists—mathematicians concerned with the precise measurement of the shape and size of the earth—with new measurements to define the best earth-fitting ellipsoid, which relates coordinates to the earth's center of mass. Unlike a local datum, an earth-centered, or geocentric, datum does not have an initial point of origin. The earth's center of mass is, in a sense, the origin. The most recently developed and widely used datum is the World Geodetic System of 1984 (WGS84). It serves as the framework for supporting locational measurement worldwide. GPS measurements are based on the WGS84 datum.

Map projection

Map projections are systematic transformations of the spheroidal shape of the earth so that the curved, three-dimensional shape of a geographic area on the earth can be represented in two dimensions, as x,y coordinates.

Maps are flat, but the surfaces they represent are curved. Transforming three-dimensional space onto a two-dimensional map is called "projection". Projection formulas are mathematical expressions that convert data from a geographical location (latitude and longitude) on a

sphere or spheroid to a representative location on a flat surface.

This process inevitably distorts at least one of these properties: shape, area, distance, or direction. For small areas, such as a city or county, the distortion will probably not be great enough to affect your map or measurements. If you're working at the national, continental, or global level, you'll want to choose a map projection that minimizes the distortion based on the requirements of your specific project.

See *Understanding Map Projections* and *Modeling Our World: The ESRI Guide to Geodatabase Design* for more on coordinate systems, datums, and map projections.

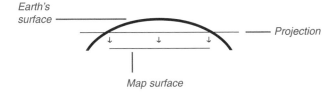

Earth's surface

Projection

Map surface

Projecting the river shapefile

The next task is to project the river shapefile into the same coordinate system as the data in the City's GreenvalleyDB geodatabase. According to your colleague at the County Water Resources Department, the river shapefile is in geographic coordinates (latitude and longitude). The rest of the data in the City's database is in the Transverse Mercator coordinate system, which is a projected coordinate system. As long as a dataset is in geographic coordinates, ArcMap can transform it on the fly to display and overlay it with the other data (as you saw in the last chapter).

However, the river data will eventually be placed into the City's GreenvalleyDB database, so you'll want to project it into the same coordinate system as the rest of the City's data to be consistent.

Projecting the shapefile is a two-step process: first you'll define the coordinate system for the shapefile; then you'll define the output coordinate system and project the file. You'll do both of these tasks in ArcToolbox. ArcToolbox contains a number of data management and conversion tools and wizards.

Define the coordinate system for the river shapefile

1. In ArcCatalog, click the Launch ArcToolbox button on the toolbar.

Launch ArcToolbox

The ArcToolbox window appears.

2. Double-click Data Management Tools in the ArcToolbox tree; double-click Projections, then double-click Define Projection Wizard. (If you are using ArcInfo you will see additional tools not shown here.)

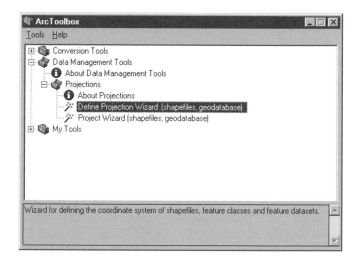

The first screen for the wizard appears.

You defined the coordinate system for the lowland shapefile using the Properties dialog box in ArcCatalog. The ArcToolbox Wizard provides an alternate way of defining a coordinate system.

3. Click the Browse button and navigate to the County_share folder under the project folder.

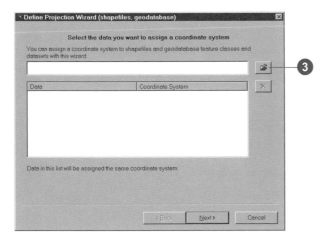

4. Click river.shp and click Add.

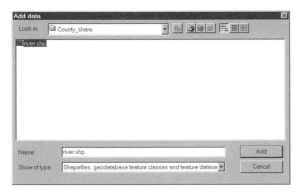

The wizard lists the shapefile. The coordinate system is listed as GCS_Assumed_Geographic_1. ArcGIS attempts to determine the shapefile's coordinate system based on the dataset's coordinate values. In this case, ArcGIS has determined that the shapefile is in geographic coordinates

(latitude–longitude); however, you need to explicitly define the geographic coordinate system before you can project the data.

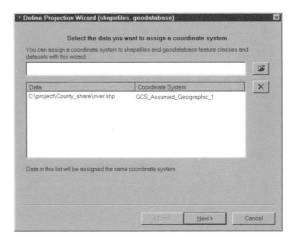

5. Click Next and click Select Coordinate System.

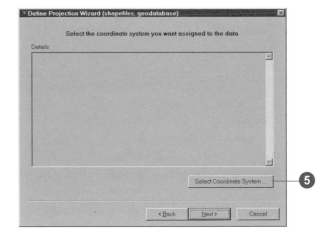

There are three ways of defining a coordinate system: using a predefined coordinate system stored as a .prj file, matching the coordinate system of an existing dataset by specifying the name of the dataset, or interactively specifying a projection and a datum and their associated parameters. In this case, you'll be specifying a predefined coordinate system.

6. Click Select on the Spatial Reference Properties dialog box.

   The wizard opens the Coordinate Systems folder.

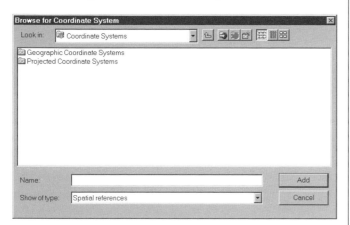

ArcGIS provides many predefined coordinate systems for you to use, stored as .prj files. The files include all the coordinate system parameters including the map projection type and parameters, measurement units, and so on. You can also define custom coordinate systems and save them as .prj files (for example, the state_dot.prj file).

7. Double-click Geographic Coordinate Systems and double-click North America.

8. Click North American Datum 1983.prj and click Add.

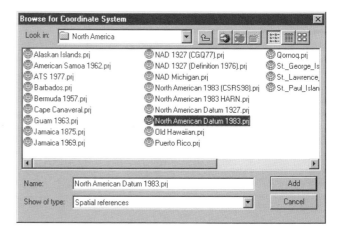

The coordinate system information is displayed in the Details window.

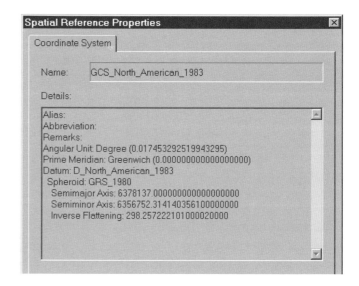

9. Click OK to close the Spatial Reference Properties dialog box, then click Next.

The wizard summarizes the coordinate system definition information.

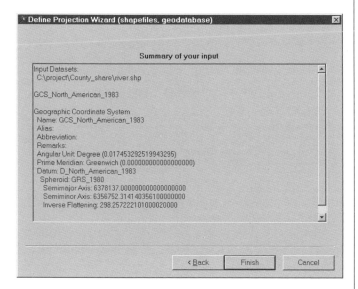

10. Click Finish.

The coordinate system of the river shapefile is now defined. To verify this, in ArcCatalog navigate to the County_share folder, click river, click the Metadata tab, and click the Spatial tab. The coordinate system is now listed as GCS_NorthAmerican_1983 (you may need to close ArcCatalog and restart it to see this).

Project the shapefile

When you define a coordinate system, you simply tell ArcGIS what projection the dataset uses and what units the coordinates are stored in. When you project a dataset, on the other hand, ArcGIS actually creates a new dataset with the coordinates transformed from the existing coordinate

units (in this case, decimal degrees) to a new coordinate system (in this case, Transverse Mercator meters). You specify the input dataset and the coordinate system to project into, and ArcGIS creates the new dataset.

Because you have data that is already in the Transverse Mercator coordinate system used by the City, you can simply specify a dataset to match to. The wizard will get the coordinate system parameters from the existing dataset and create a new river shapefile in that coordinate system.

1. In ArcToolbox, double-click Project Wizard.

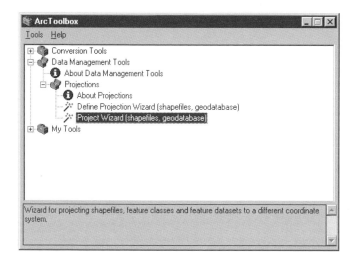

2. Click the Browse button on the wizard and navigate to the County_share folder under the project folder.

3. Click river.shp and click Add.

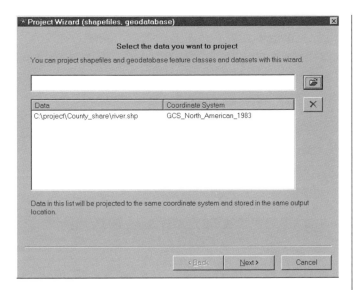

The shapefile name and coordinate system you defined appear in the window.

4. Click Next.

The wizard asks you to specify a name for the projected shapefile and a location where it will be stored. You'll put it in the City_share folder since it will eventually become part of the City's database. You'll name it river02prj since it will be the second version of the river dataset and will have been projected.

5. Click the Browse button and navigate to the project folder. Double-click City_share, then type river02prj in the Name text box.

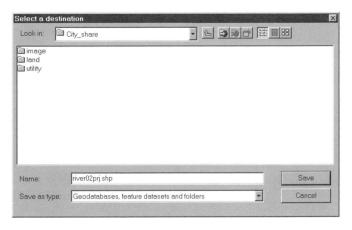

6. Click Save.

The City_share folder is listed as the location to store the projected shapefile river02prj.

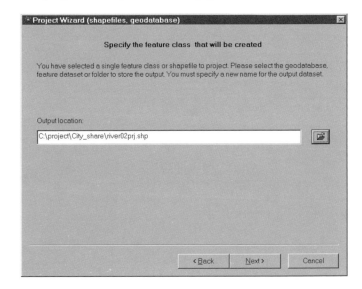

7. Click Next.

The wizard asks you for the coordinate system into which to project the river shapefile.

8. Click Select Coordinate System.

The Spatial Reference Properties dialog box appears.

This is the same dialog box you used to define the coordinate system for the lowland shapefile and the river shapefile. In those cases you specified a spatial reference (.prj) file. This time you'll specify an existing dataset from which to get the coordinate information. You know the parks feature class is in the right coordinate system since you copied it directly from the City's existing geodatabase.

9. Click Import and navigate to the WaterProject geodatabase under the project folder connection.

10. Click parks_polygon and click Add.

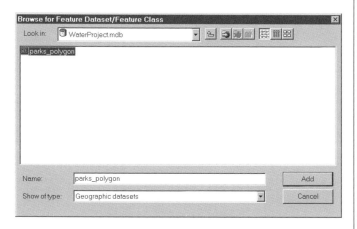

The dialog box displays the coordinate system, and you can see it's the correct one: PCS_Transverse_Mercator (PCS stands for Projected Coordinate System).

11. Click OK to close the dialog box.

The wizard presents a summary of the output coordinate system parameters.

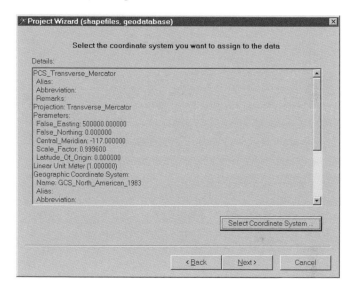

12. Click Next.

The wizard shows you the output extent of the projected file.

13. Click Next, then click Finish.

The Project Wizard projects the river shapefile to match the coordinate system of the data in the City's database. The projected shapefile, river02prj, is saved in the City_share folder.

14. You're finished using ArcToolbox, so go ahead and close it by clicking the x in the upper-right corner of the window.

Exporting the river shapefile to the geodatabase

The projected river dataset will eventually go into the City's geodatabase. You'll export the dataset to a feature class in the WaterProject geodatabase now, so it will be in the right format to be copied into the City's geodatabase later.

1. In the Catalog tree, navigate to the City_share folder, right-click river02prj, point to Export, and click Shapefile to Geodatabase. (If you are using ArcInfo, you will see additional export options not shown here.)

The Shapefile to Geodatabase dialog box appears. The name of the input shapefile is already filled in.

2. Click the Browse button next to the Output Geodatabase text box and navigate to the project folder.

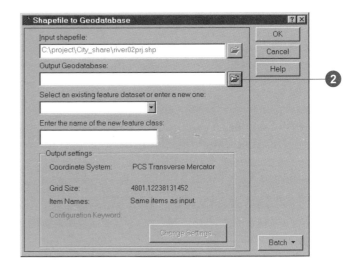

3. Click WaterProject.mdb and click Open.

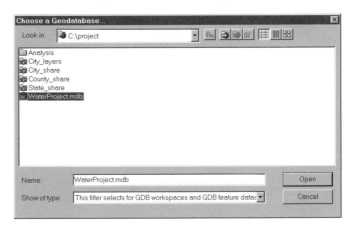

The WaterProject database is added to the dialog box as the output geodatabase.

4. Click in the feature class name text box and type "river03exp" to indicate that this is the third version of the river and it has been exported to the geodatabase.

You don't need to specify a feature dataset since you are creating a standalone feature class.

5. Click OK.

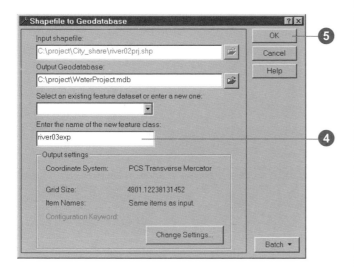

ArcGIS shows the progress of the export operation. When it finishes, navigate to the WaterProject geodatabase and double-click to display its contents. The river03exp feature class is listed (you may need to refresh the tree by clicking View and clicking Refresh to see it).

The next task is to update the parks layer with the new historic park.

Digitizing the historic park

You need to add the new historic park to the parks layer so you can include it in the buffer around parks you'll create during the analysis. The Parks Department has not yet added the planned Homestead Historic Park to the parks feature class in the City's database, though they have settled on the boundary. You'll digitize the park boundary from a scanned image of the boundary drawn on a map.

You'll digitize the new park into the copy of the parks feature class in the WaterProject geodatabase. The updated file will replace the original file in the City's database later, after the Parks Department checks to make sure the new park was added correctly.

After opening a new map in which to do the digitizing, you'll register the scanned image to the streets layer. You'll then digitize the park boundary and add the attributes for the new park.

Open a new map

You'll digitize the new park in a new map in ArcMap. You'll need to add four datasets to the map: the existing parks feature class that you'll be adding the new park to; the scanned image of the park boundary (stored as a TIFF file), which you'll use as a guide while digitizing; the streets layer that you'll use to register the scanned image; and the parcel_2 coverage that you'll use to snap the park boundary to since the boundary aligns with the parcel boundaries.

1. Click the Launch ArcMap button on the ArcCatalog toolbar.

If the ArcMap startup dialog box appears, click the option for A new empty map and click OK.

2. In ArcCatalog, navigate to the WaterProject geodatabase under the project folder connection and open it by double-clicking it or clicking the plus sign next to it, if necessary.

3. Click and drag parks_polygon onto the map in ArcMap.

4. Add the parcel_2 coverage to the map by opening the City_share\land folder in ArcCatalog, clicking parcel_2, and dragging it onto the map.

The parks_polygon and parcel_2 layers are displayed on the map.

5. Click the Full Extent button on the Tools toolbar to see all of both layers.

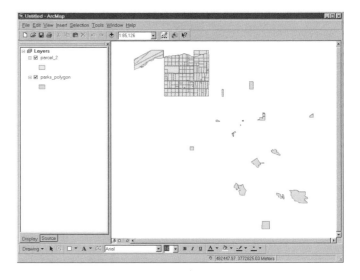

6. Add the streets layer to the map by opening the City_layers folder, clicking streets, and dragging it onto the map.

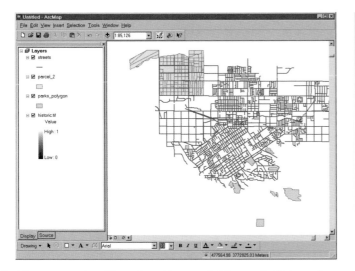

7. Now open the City_share\image folder and click and drag historic onto the map.

You get a warning that the layer is missing spatial reference information—its coordinate system is undefined.

That's OK because you'll be registering the image interactively to place it in the correct geographic space. Click OK to close the warning message.

You'll notice that the image is not displayed even though it has been added to the map. That's because it's in scanner units (inches) rather than in geographic coordinates.

8. Click the Full Extent button on the Tools toolbar.

When the map redraws, you can see the City data layers as a small dot in the upper center of the view. The scanned image is actually drawn in the lower center, but it's too small to even see. You saw a similar situation in Chapter 5, 'Assembling the database', when you first added the lowland shapefile to the map.

The extent of the image, which is in inches, is 0 to about 13 in both the x and y directions. The extent of the other data, which is actual geographic space represented in UTM meters, is from about 478,000 to about 490,000 in the x direction and about 3,765,000 to 3,772,000 in the y direction. ArcMap attempts to draw all the data on the same page, so the page extent ranges from values of 0,0 in the lower left of the page to values of over 490,000 in the x direction and over 3,772,000 in the y direction. The image and the data are drawn in their respective parts of the page and end up being very small. You need to register the image so it is in the same geographic space as the streets, parcels, and parks.

9. Right-click historic.tif in the ArcMap table of contents and click Zoom To Layer.

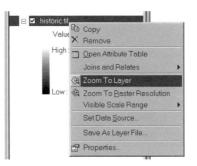

Now you can see the image.

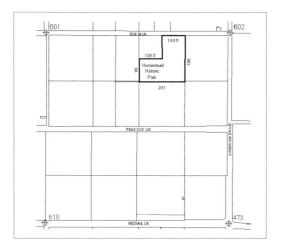

Before registering the image, save the map in case you need to stop or take a break during the process.

10. Click File and click Save.

11. Navigate to the project folder. Name the map "water project" and click Save.

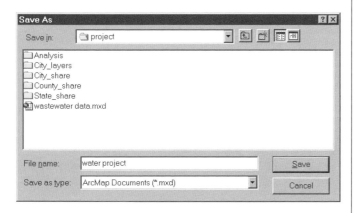

You'll use this map for the rest of the project.

Find the park area on the streets layer

The Parks Department added several registration marks to the sketch at street intersections. You'll register the image to the streets by interactively adding links between the image and the streets layer—first pointing at a registration mark on the image, then at the corresponding intersection on the streets layer. You'll need to find the area on the streets layer corresponding to the area covered by the image. To make it easier to do this, you'll first open an overview window, so you'll be able to see the image and the streets at the same time.

1. Click Window and click Overview.

A small window appears showing the image.

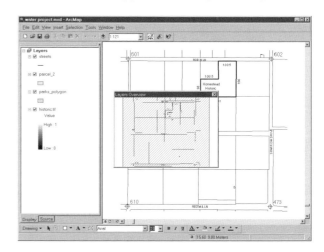

The overview window displays whichever layers are displayed in the main window at the time you open it (although you can modify this once the window is created). By default, it also shades the area visible on the main screen. (Since the main window and overview window show the same area at this point, the shading covers the entire image in the overview window.) When you zoom to the streets, the overview window will continue to display the image, so you'll be able to see both.

Now you can find the park area on the streets layer and zoom to it.

The image has several streets marked on it in the vicinity of the park including Robin Lane, Peacock Lane, and Sparrow Drive. You can search for one of these on the streets layer to find the area the park is in.

2. Click the Edit menu and click Find.

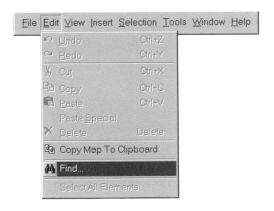

3. In the Find text box type "Peacock".

4. Click the In layers dropdown arrow, scroll down, and click streets.

5. Click In fields, click the dropdown arrow, and click NAME.

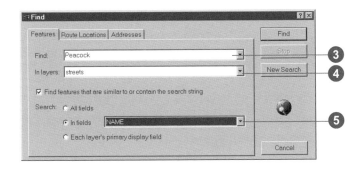

6. Click Find.

   Peacock appears in the list of features that have been found.

7. Right-click Peacock and click Zoom to feature(s).

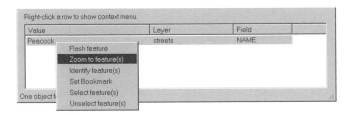

The map displays the area that includes Peacock Lane. The overview window still displays the image. You may want to enlarge the overview window so it's easier to see the image—simply click and drag one of the corners of the window. You may also want to move the overview window to make it easier to see the streets.

8. Click Cancel to close the Find dialog box.

To make sure you're in the right area, label the streets.

9. In the table of contents, right-click streets and click Label Features.

You can see that you're in the area that includes the park.

10. Use the Zoom In tool on the Tools toolbar to draw a box around the four street intersections that correspond to the control points on the scanned image. Use the image displayed in the overview window to orient yourself.

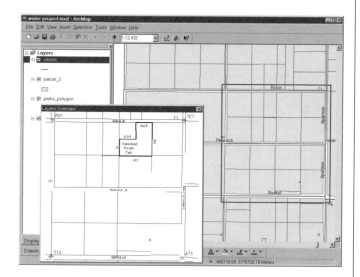

11. Close the overview window.

Now the display is zoomed to the area corresponding to the image.

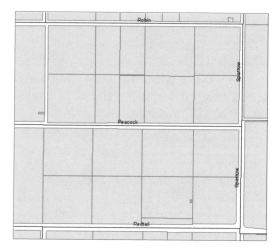

12. You won't need to display parcel_2 for the next set of steps, so uncheck the layer to make it easier to see the streets.

Register the scanned image

Now you're ready to register the image to the streets layer. You'll do this by adding links between the control points on the image and the corresponding street intersections on the streets layer. This is known as georeferencing. ArcMap requires a minimum of three links to transform the image—rotating, scaling, and warping it as needed to fully register it to the streets.

1. Click the View menu, point to Toolbars, and click Georeferencing.

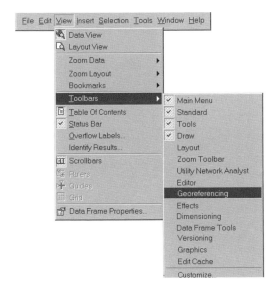

The Georeferencing toolbar appears.

2. Click the Layer dropdown arrow and click historic.tif.

3. Click the Georeferencing dropdown arrow and click Fit To Display.

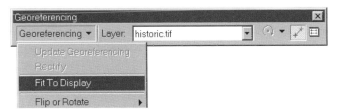

ArcMap scales the image to fit in the current window.

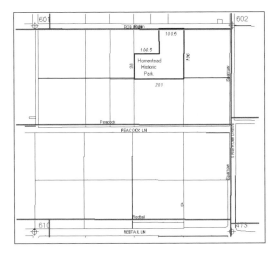

Since the window is currently zoomed to the four street intersections that match the registration marks on the image, the streets and the image are displayed at approximately the same scale. You can see, though, that the control points aren't located exactly at the intersections. You'll add three links to register the image. To make it easier, you'll use a magnifier window. You can add control points within the magnifier window.

4. Click Window and click Magnifier.

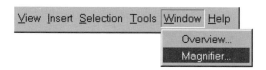

A small window appears with a default magnification of 400 percent.

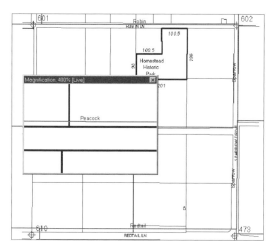

When you drag the window and release the mouse button, it magnifies the portion of the display it is currently on top of by 400 percent.

5. Click the Add Control Points button on the Georeferencing toolbar.

Add Control Points

The cursor turns into a crosshair.

6. Drag and center the magnifier window over the registration mark in the upper right, labeled 602, and release the mouse button. If necessary, reposition the window so you can see both the registration mark and the corresponding street intersection (Robin and Sparrow) within the window.

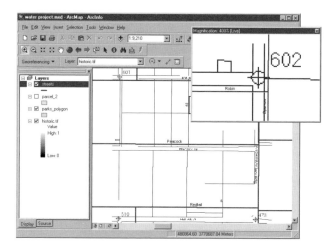

Note that the registration mark and intersection may be in slightly different positions on your map.

7. Center the cursor over the registration mark and click.

A green control point is added to the image. Move the cursor away from the control point but don't click again.

A line stretches from the control point as you move the cursor. This is the link—you'll connect the other end of it to the corresponding street intersection.

8. Center the cursor over the intersection of Robin and Sparrow on the streets layer (you can see the link stretch as you do this) and click.

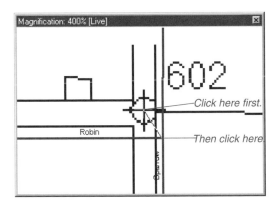

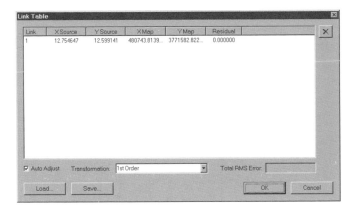

You've added the other end of the link. The second control point is shown as a red crosshair. The image is now shifted so the upper-right corner of the image is in the right location (at the corner of Robin and Sparrow). However, not all registration marks are located exactly at the intersections. You'll add a couple more links to achieve better registration.

Before you add the next link, look at the link table.

9. Click the View Link Table button on the Georeferencing toolbar.

View Link Table

For each link, the table lists the x- and y-coordinates for the source (the scanned image) and the corresponding coordinates for the map (the streets layer).

If you make a mistake and need to delete a link, select it and click the Delete button, which looks like the letter x.

10. Click Cancel to close the Link Table.

Now you'll add the other two links.

11. Drag and center the magnifier window over the registration mark in the upper left, labeled 601, and release the mouse button.

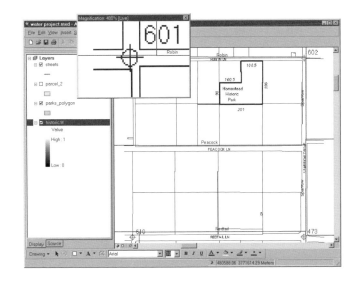

If necessary, reposition the window so you can see both the registration mark on the image and the corresponding street intersection.

12. Center the cursor over the registration mark and click.

13. Center the cursor over the intersection and click to add the second control point.

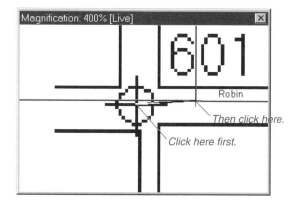

The image adjusts slightly. Now add the third link in the same manner.

14. Drag the magnifier window over the registration mark in the lower right, labeled 473.

15. Click on the registration mark, then click on the intersection.

The image shifts again.

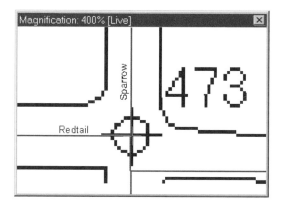

Now the control points match up pretty closely with the intersections. You could add a fourth link, but at this point the registration is acceptable for your purposes.

16. Click the Georeferencing dropdown arrow and click Update Georeferencing to save the new registration.

You don't need the control points anymore.

17. Click the Georeferencing dropdown arrow and click Delete Control Points. Then close the Georeferencing toolbar.

Keep the magnifier window open since you may want to use it when digitizing the park boundary.

Display the park boundary and the parcels

You need to be able to see the image underneath the parcels when digitizing, so you'll draw the parcel outlines.

1. Click the legend symbol under parcel_2 in the table of contents.

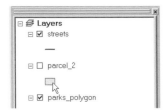

2. In the Options panel on the Symbol Selector dialog box, click the Fill Color dropdown arrow and click No Color.

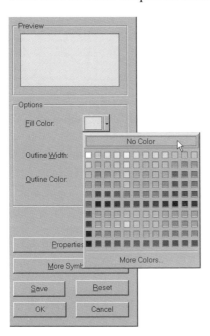

3. Click the Outline Color dropdown arrow and set the color to red.

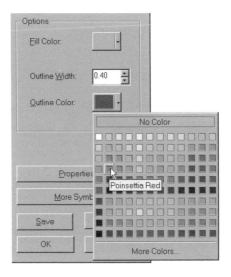

4. Click OK.

5. Check the box next to the parcel_2 layer to draw the parcels.

You don't need to display the streets or the street names at this point, but you'll keep the layer on the map for use in the analysis phase of the project.

6. Right-click streets in the table of contents and click Label Features (it's currently checked) to turn off the street name labels.

7. Uncheck the streets layer so the streets are no longer displayed.

Now your display should show only the parcel boundaries in red on top of the image.

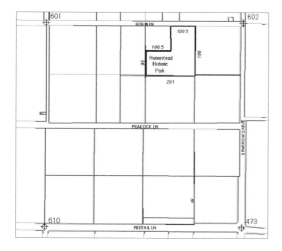

8. Click File and click Save to save your map display.

You can see that the parcel boundaries in the parcel_2 layer match up pretty well—but not exactly—with the parcel boundaries on the image. Since you'll be snapping the boundary to the parcel_2 layer and using the image merely as a guide for digitizing the park boundary, this registration is acceptable. If you were digitizing solely by tracing from the image, you would want to make sure the image was as closely registered to the streets as possible. To do this, you could add more links. In any case, the registration might never be exact because of factors such as distortion on the scanned image, the placement of registration marks on the image, and slight differences in the location of the streets on the image (created from a paper map) and in the GIS database.

Prepare to digitize the park boundary

You'll align the park boundary exactly with the surveyed parcel boundaries by snapping to the parcel_2 layer. First you need to set up the digitizing environment.

1. Zoom in to the park by using the Zoom In tool on the Tools toolbar to draw a box around the park boundary.

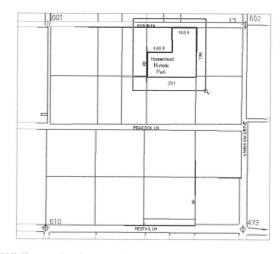

While you're here, add a bookmark to use when you start digitizing the boundary.

2. Click the View menu, point to Bookmarks, and click Create. Name the bookmark Park Boundary and click OK.

3. Click the Editor Toolbar button.

Editor Toolbar

4. Click Editor and click Start Editing.

The Start Editing dialog box appears. You'll be adding a feature to the parks polygon feature class you copied to the WaterProject geodatabase, so select it as the database to edit data from.

5. Click project\WaterProject.mdb, then click OK.

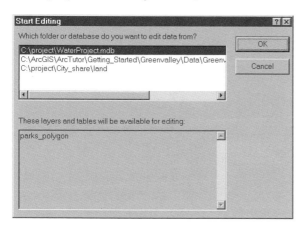

The Editor toolbar shows that the target layer (the one you are editing) is parks_polygon and the current editing task is Create New Feature.

Now set the snapping environment so the new park boundary will align exactly with the existing parcel boundaries.

6. Click Editor, then click Snapping.

7. Check the box in the Vertex column for the parcel_2 layer.

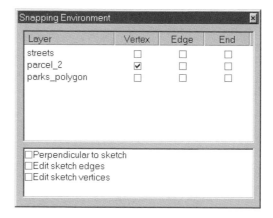

This will snap the editing cursor to the vertices of the parcels.

8. Close the Snapping Environment dialog box.

The snapping tolerance defines how close the cursor must be to an object before it snaps to that object. You can change the snapping tolerance by choosing Options from the Editor menu. For this exercise you don't need to change the snapping tolerance.

Start digitizing the boundary

1. Click the Create New Feature tool.

Create New Feature

It may help you to zoom in to the northeast corner of the park—you can use the Pan and Zoom tools while you are digitizing. Just click the Zoom tool, drag a rectangle around the area you want to see, and click the Create New Feature tool again to resume digitizing. Then use the bookmark you created to display the full park boundary again. Or, use the magnifier window. You can move the window—and move the cursor in and out of the window—while digitizing.

If you make a mistake while digitizing, click the Undo button on the ArcMap Standard toolbar.

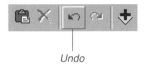

Undo

2. Move the editing cursor over the northeast corner of the Homestead Historic Park boundary.

The existing parcel boundaries are drawn with red lines, and the park boundary is drawn with a heavy black line. You'll snap to the existing parcel boundary. The digitizing cursor is shown as a blue dot with a crosshair. When the cursor gets within the snapping tolerance of the parcel corner, the blue dot snaps to the vertex.

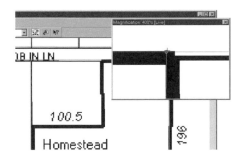

3. Click the northeast corner of the park to start your edit sketch.

4. Move the cursor to the southeast corner of the park. There are two vertices here. Make sure that the cursor snaps to the southernmost vertex. Click the vertex.

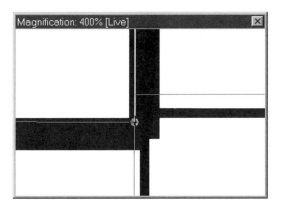

After you add the second point, a line is drawn back to the first point in your sketch to create a polygon that stretches as you move the cursor.

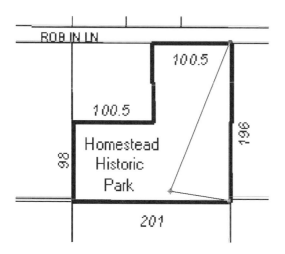

5. Move the cursor to the southwest corner of the park. There are two vertices here. Click the southernmost vertex.

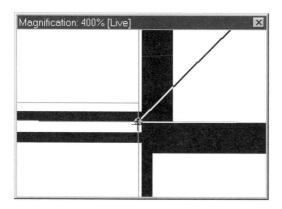

Place a vertex by angle and distance

The next segment of the park boundary is only half as long as the parcel boundary. The scanned image includes dimensions for each segment. You'll use the dimensions to place the next two vertices using angle and distance.

1. Place the cursor near the parcel boundary line at the corner of the park. Right-click and click Parallel.

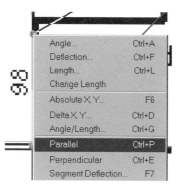

Now the cursor is constrained to be parallel to the parcel boundary.

2. Right-click again and click Angle/Length.

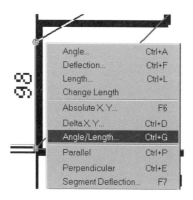

The Angle/Length dialog box appears.

3. Click in the Length text box and type "98". Press Enter.

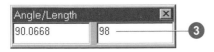

The length of this segment of the park boundary as shown on the image is 98 meters.

A vertex is placed on the line coinciding with the parcel boundary 98 meters from the previous vertex you added at the corner of the parcel.

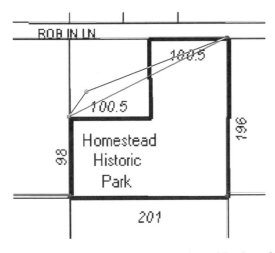

4. Move the editing cursor near the midpoint of the north boundary of the parcel. Right-click and click Parallel.

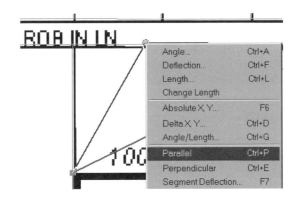

Now the next segment you add will be parallel to the north boundary of the parcel.

5. Right-click again and click Angle/Length.

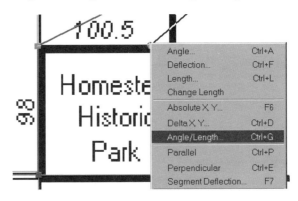

6. Click in the Length box, type "100.5" as the length, and press Enter.

The new segment is added to the edit sketch; it is 100.5 meters long and parallel to the north boundary of the parcel.

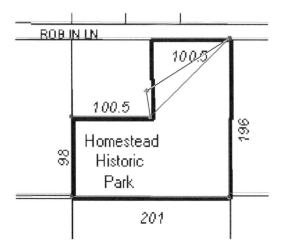

Add a perpendicular line

The next segment is perpendicular to the segment you just added.

1. Move the cursor along the vertical line, partway to the north boundary of the parcel, right-click, and click Perpendicular.

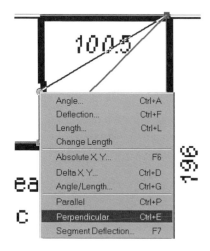

Now the cursor is constrained to be perpendicular to the previous segment.

2. Click to add a vertex partway to the north boundary.

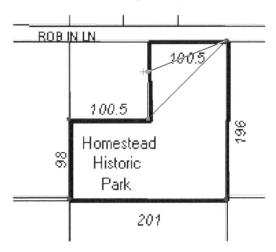

Add a point at the intersection of lines

1. Click the Create New Feature tool dropdown arrow and click the Intersection tool.

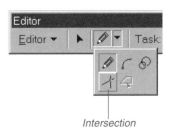

Intersection

The Intersection tool lets you place the next point in your edit sketch at the intersection of two lines.

2. Place the cursor near the segment you just created. A temporary line extends along the segment. You will make this the first intersection line. Click to set the intersection line.

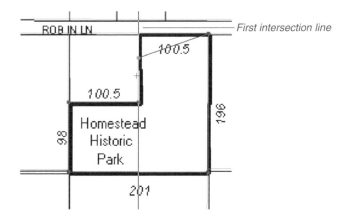

First intersection line

3. Place the cursor near the north boundary of the parcel. You will make this the second intersection line. Click to set the intersection line.

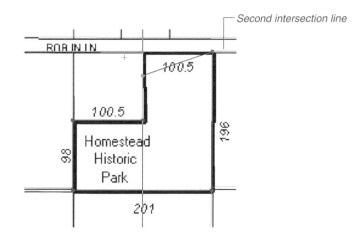

Second intersection line

The new vertex appears at the intersection of the two lines at the corner of the boundary.

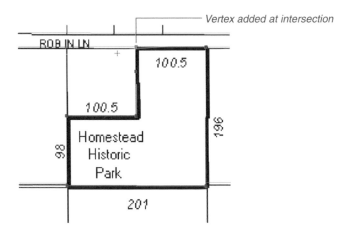

Vertex added at intersection

Finish digitizing

1. Click the northeast corner of the park, right-click, and then click Finish Sketch.

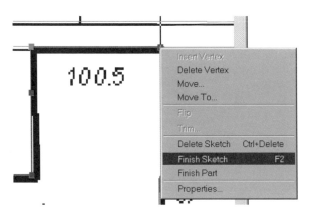

The new park polygon is finished. Its boundary turns light blue to indicate that it is selected, and it takes on the color of the other park polygons.

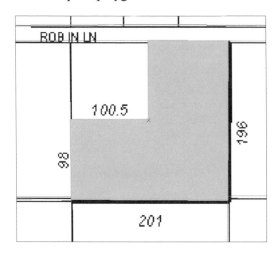

2. Close the magnifier window if it's still open.

Edit the feature attributes

Now that you've finished digitizing the park, you can update the new feature's attributes in the database.

1. Click the Attributes button on the Editor toolbar.

Attributes

2. Click next to Name in the Value column and type "Homestead Historic".

The park will be maintained by the City, so you will update the maintenance field as well.

3. Type "City" as the value for the Maintenance field and press Enter. Close the Attributes dialog box.

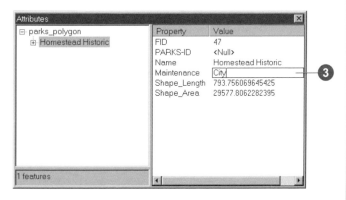

4. Right-click parks_polygon in the table of contents and click Label Features.

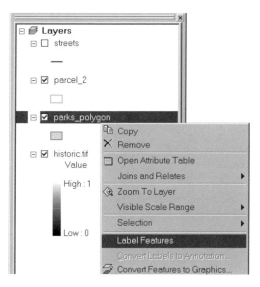

The new park is labeled with its name.

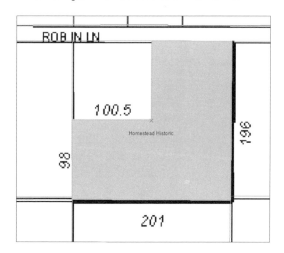

Save your edits

Now save your changes to the park feature class in the database.

1. Click Editor, then click Save Edits.

2. Click Editor, then click Stop Editing.

Although this was a simple example, you saw how the Editor provides a variety of tools for constructing features.

Before continuing, clean up the table of contents. You don't need the historic.tif layer anymore, so remove it from the map.

3. Right-click historic.tif and click Remove.

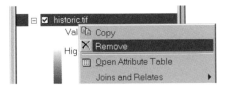

4. You don't need to display parks right now, so uncheck them.

5. Click File and click Save to save your map at this point.

The last task before you start the analysis is to merge the two parcel layers.

Merging the parcel layers

Sometimes data that you need is in two or more adjacent datasets, either because of the way the data was created or the way it is stored. For example, multiple datasets might be created by digitizing from adjacent map sheets into separate shapefiles. In some cases, data for large areas is stored as sets of separate tiles to make it easier to manage and update the features.

The parcel data that you will use in the analysis is stored as tiles of adjacent shapefiles. You have two tiles that you will merge into a single layer so you'll be able to more easily select the suitable parcels during the analysis.

First display the area covered by the parcels.

1. Right-click parcel_2 and click Zoom To Layer.

Now you'll add the other parcel layer to the map.

2. Click the Add Data button, navigate to the City_share\land folder under the project folder, click parcel_1, and click Add.

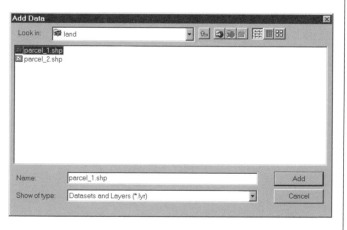

You can see the two adjacent parcel layers. Parcel_1 is drawn with a default solid fill color, while the parcel_2 boundaries are displayed in red.

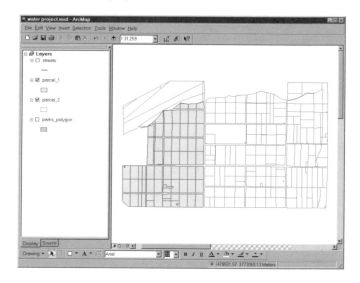

3. Click the Tools menu and click GeoProcessing Wizard.

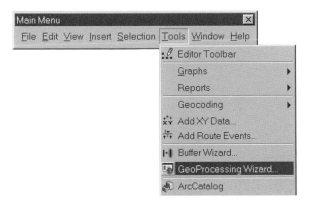

The GeoProcessing Wizard lets you combine features and datasets in several ways. In this case you'll be merging two datasets.

4. Choose Merge layers together and click Next.

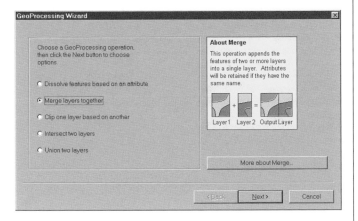

5. Select parcel_1 and parcel_2 as the layers to merge by checking the boxes.

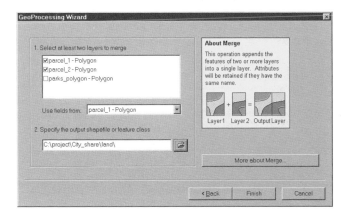

You'll put the joined layer in the Analysis folder, where you'll store the other analysis layers.

6. Click the Browse button next to the "Specify the output" text box and navigate to the project\Analysis folder.

7. In the Name text box, type "parcel01mrg", then click Save.

8. Click Finish to merge the parcels.

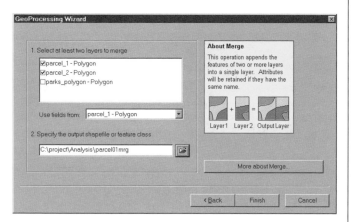

Now all the parcels are in a single layer.

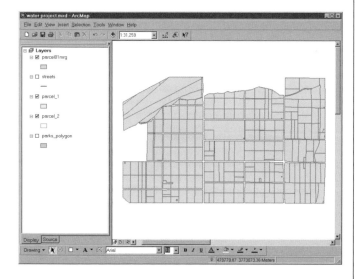

9. You don't need the parcel_1 and parcel_2 layers any longer, so remove them from the map by right-clicking them and clicking Remove.

10. Click File and click Save to save your map at this point.

In this chapter you have prepared your data for the analysis and completed the project database. Preparing data—whether by converting it, changing its coordinate system, managing its attributes, or editing its features—is a very important part of a GIS project. The quality of the analysis and map depends on the quality of the data. An organization's biggest investment in a GIS project, other than its staff, is usually in its data.

In the next chapter you'll perform the analysis to find parcels that meet the City's criteria as a site for the wastewater treatment plant.

Performing the analysis

7

In the planning phase you decided which data you would need in order to meet the criteria. You then assembled the data and prepared it for analysis. At this point you're ready to perform the analysis.

There are often several alternative methods you can use to get to the end result. The approach you use is partly dictated by the problem you're trying to solve, partly by the data you're using, and partly by personal preference.

In this analysis, the parcels suitable for a site for the wastewater treatment plant have to meet certain location criteria and also have to be vacant.

The location criteria fall into two categories: features the plant should be away from or outside of (away from parks and residences, outside the flood zone) and features the plant should be near or inside of (near the river, inside the low elevation area). The areas delineated by these criteria become the unacceptable and acceptable areas for the plant, respectively.

You'll find the parcels that fall outside the unacceptable areas, then find the subset of these that fall inside the acceptable areas. Next, of the remaining parcels, you'll find the ones that are vacant.

The City also prefers that the site be near a road and near the existing wastewater junction. You'll find and tag the parcels that are within 50 meters of a road. You will also find and tag the parcels that are within 500 and 1,000 meters of the wastewater junction. Finally, you'll find parcels that are at least 150,000 square meters in size, large enough to build the plant on.

Setting up for analysis

If you closed ArcCatalog and ArcMap at the end of Chapter 6, 'Preparing data for analysis', you'll need to reopen them and reopen the water project map. Your map should currently include the parks_polygon, streets, and parcel01mrg layers, with the parcels displayed.

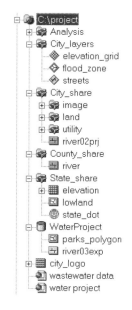

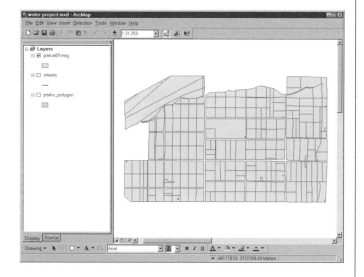

You'll be adding data from several locations during the analysis, so in the Catalog tree navigate to and open the project folder so you can see the City_layers, City_share, and State_share subfolders and the WaterProject geodatabase. Open each subfolder and the geodatabase so you can see the contents of each.

Now you're ready to start the analysis process. The general approach for this analysis was presented in the flowchart in Chapter 4, 'Planning a GIS project'. At this point, you need to develop the specific steps for each phase of the analysis. Detailed flowcharts for each section are presented at the beginning of the appropriate section in this chapter. You'll implement the analysis using ArcMap.

Combining the river buffer and lowland layer and combining the park and residential buffers with the flood zone layer will reduce the number of selections you'll have to do to find the parcels that meet the criteria. It's easier to do the selections all at one time, so you'll first create the two layers that delineate the acceptable and unacceptable areas, then do all three selection steps.

Delineating the area the plant site should be within

In this phase of the analysis you'll buffer and combine features to delineate the areas the wastewater treatment plant should be within (areas within 1,000 meters of the river, and lowland areas). You'll create a 1,000-meter buffer around the river, then combine the buffer with the lowland layer. Here are the steps:

- Create a 1,000-meter buffer around the river.

- Overlay the river buffer and the lowland layer.

Here's a flowchart of the process using the layer names:

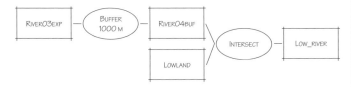

Buffer the river

You'll use the Buffer Wizard in ArcMap to buffer river03exp to 1,000 m to create river04buf.

1. Uncheck the parcel01mrg layer since you won't need to display it until later.

2. In the Catalog tree, click riv03exp in the WaterProject geodatabase and drag it onto the map in ArcMap.

3. Click the Full Extent button on the Tools toolbar to see the river.

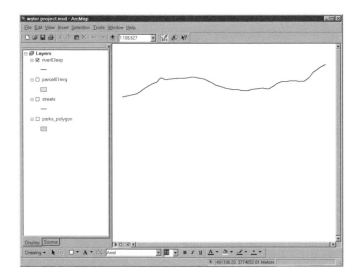

4. Click the Tools menu and click Buffer Wizard.

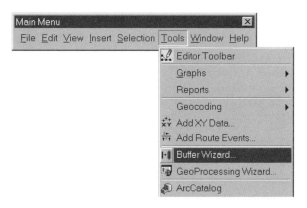

5. Click the option to buffer the features of a layer, click the dropdown arrow, and click river03exp as the layer to buffer. Click Next to display the next screen.

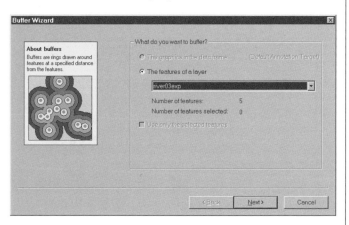

The Buffer Wizard gives you options for specifying the buffer distance and for the number of buffers to create.

6. Click the option to buffer at a specified distance and type 1000 in the text box. Then click the Distance units dropdown arrow in the bottom panel and click meters to create a buffer of 1,000 meters around the river. Click Next.

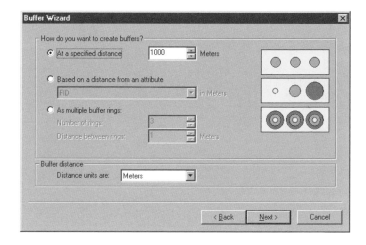

7. Click Yes to dissolve the barriers between buffers (the river is actually composed of five line segments, each of which will be buffered, so dissolving the barriers will create a single buffer around the river).

8. In the bottom panel, click the option to save the buffer in a new layer, type the path to the Analysis folder, and type river04buf as the layer name.

For the rest of the analysis, the output layers you create will be shapefiles in the Analysis folder.

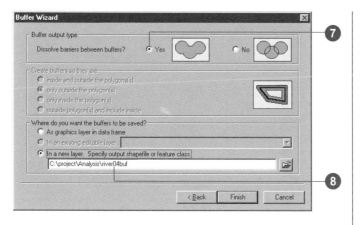

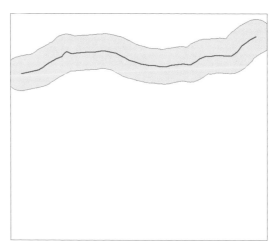

9. Click Finish.

   ArcMap creates the buffer and displays it.

10. Drag the river03exp layer above the river04buf layer in the table of contents to display it on top.

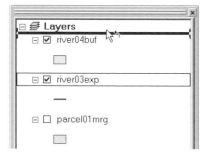

Overlay the river buffer and the lowland area

Next you'll use the GeoProcessing Wizard to combine the lowland layer and the river buffer to create low_river, which delineates the area the plant site needs to be within.

1. Add the lowland shapefile to the map by dragging it from the State_share folder in the Catalog tree.

ArcMap warns you that the lowland shapefile is in a different coordinate system than the other data on the map. That's okay—since you've defined the coordinate system for lowland (in Chapter 6, 'Preparing data for analysis'), it will overlay correctly with the other data. Click OK to close the Warning message box.

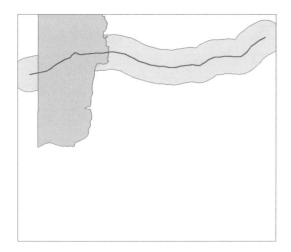

2. Click the Tools menu and click GeoProcessing Wizard.

3. Click the option to Intersect two layers.

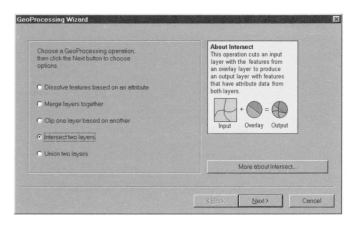

Intersect creates a layer containing only the area covered by both input layers.

4. Click Next.

5. In the first box, click the dropdown arrow and click lowland as the input layer.

6. In the second box, click the dropdown arrow and click river04buf as the overlay layer.

7. In the third box, make sure the path to the Analysis folder is displayed and type low_river as the layer name. If the path to the Analysis folder is not displayed, type it in or use the Browse button to navigate to it, then type the name of the new layer.

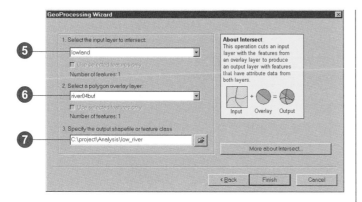

9. Uncheck low_river, river03exp, lowland, and river04buf since you don't need to display them for the next step. (Keep them on the map since you may want to use them later to verify the results of your analysis.)

10. Click File and click Save to save your map so far.

Through the rest of the chapter, you'll want to periodically save your map in case you need to stop or take a break and come back to it. We'll remind you at the end of each section, but you may want to save more often.

8. Click Finish.

The new layer is displayed. It contains only the area covered by both the river buffer and the lowland layer. You can see the lowland and river buffer layers underneath and that the new layer, low_river, represents the intersection of the two.

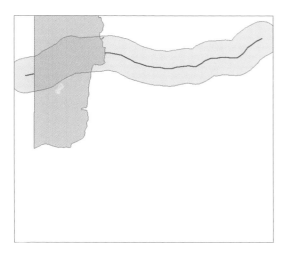

Delineating the areas the plant site should be outside of

Now you'll find the areas the plant should be outside of (areas within 150 meters of parks and residential parcels and within the flood zone). You'll create a 150-meter buffer around parks. Then select the residential parcels from the parcel layer and create a 150-meter buffer around them. You'll then combine the two buffer layers and combine the result with the flood zone layer. The resulting layer will delineate the areas the plant site should not be within. Here's a flowchart of the process:

Here are the steps:

- Buffer the parks to 150 meters.

- Select the residential parcels.

- Buffer the residential parcels to 150 meters.

- Overlay the parks and residential buffers.

- Overlay the combined park and residential buffer layer with the flood zone layer.

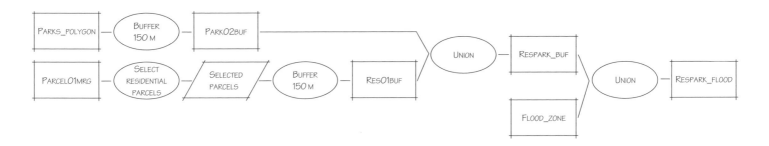

Buffer the parks

First you'll create a 150-meter buffer around parks.

1. Check parks_polygon to display the parks.

   The parks are labeled with their names.

2. Right-click parks_polygon and click Label Features (which is currently checked on) to turn off the labels.

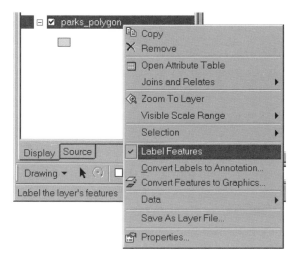

3. Click the Tools menu and click Buffer Wizard.

4. Click the dropdown arrow and click parks_polygon as the layer to buffer. Then click Next.

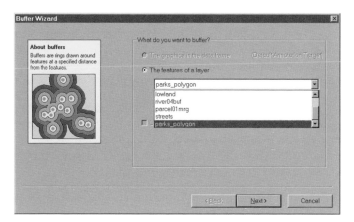

5. Click the option to buffer at a specified distance, type 150 as the buffer distance, and click Next.

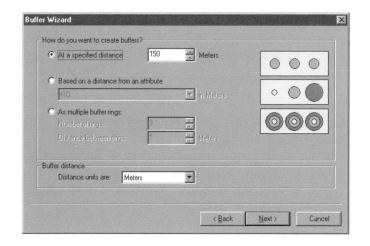

6. On the next screen, click Yes for the option to dissolve between buffers.

7. Click the option to create buffers only outside the polygon(s).

ArcMap gives you several options for creating buffers inside or outside the boundaries of polygons (this isn't an option when buffering points and lines since there is no "inside"). In this case you're only interested in finding the distance outward from the edge of each park.

8. Click the option to create the buffers in a new layer, make sure the path to the Analysis folder is specified, and type park02buf as the layer name. Then click Finish.

The park buffers are displayed.

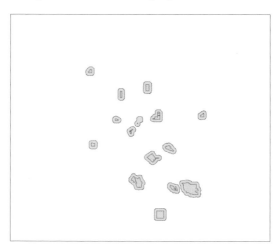

Select the residential parcels

Previously, you created buffers for a layer with a single feature (the river) and a layer with many features (parks). This time you'll buffer selected features in a layer—only the residential parcels from the parcel01mrg layer. You need to know the land use code for residential parcels so you can select them. You can get the code from the metadata.

1. In ArcCatalog, click parcel_1 in the City_share\land folder and click the Metadata tab.

2. Click the Stylesheet dropdown arrow and click FGDC FAQ.

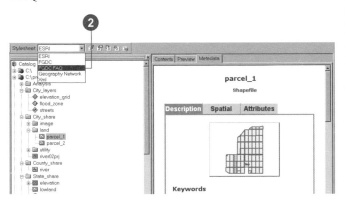

The format of the metadata changes.

The metadata that you see displayed in Catalog is controlled by the stylesheet that you use to display it. The stylesheets are similar to database queries—they basically define what information is pulled out of the metadata and how it can be formatted.

ArcGIS provides several predefined stylesheets—the default is the ESRI stylesheet, which you used earlier. You can also create your own stylesheets.

The FGDC FAQ stylesheet was developed by the Federal Geographic Data Committee to present the metadata in the form of a set of frequently asked questions. This format lets you see the values for each attribute in a layer (as long as they've been defined in the metadata).

3. Click "7. How does the data set describe geographic features?" in the first section.

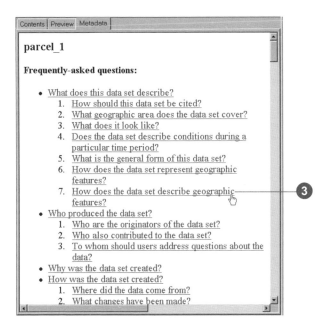

The value definitions for the land use attribute are listed (you may need to scroll down to see them). You can see that residential parcels have a value of 510. (Note also that vacant parcels have values of 713, 723, and 732. You'll use these values later in the analysis.)

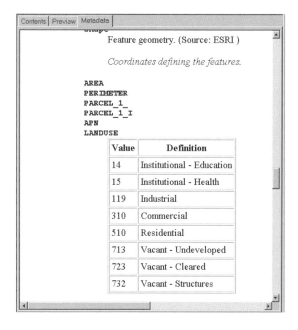

Value	Definition
14	Institutional - Education
15	Institutional - Health
119	Industrial
310	Commercial
510	Residential
713	Vacant - Undeveloped
723	Vacant - Cleared
732	Vacant - Structures

Before selecting the residential parcels, zoom in to the parcel layer.

4. Right-click parcel01mrg in the ArcMap table of contents and click Zoom To Layer, then check the layer to display the parcels.

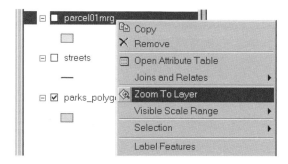

You can also see the buffer around the historic park.

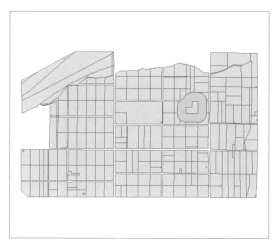

5. Click the Selection menu and click Select By Attributes.

6. In the Select By Attributes dialog box, click the dropdown arrow next to the Layer box and click parcel01mrg as the layer to select from.

The default method is to Create a new selection, which is what you want.

You'll use the query builder to create a simple query expression.

7. Double-click LANDUSE in the Fields list.

8. Click the equals sign (=) button.

9. Double-click 510 (the code for residential) in the Unique values list.

You can see the query expression you've built in the text box. It should look like this:

"LANDUSE" = 510

10. Click Apply. The residential parcels are highlighted with a blue line. Close the Select By Attributes dialog box.

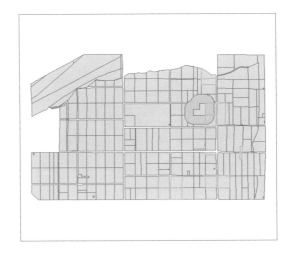

Now you'll be able to create buffers around the parcels to keep the plant from being sited too close to a residence.

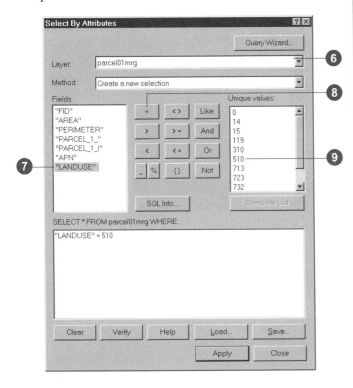

Buffer the selected parcels

1. Click the Tools menu and click Buffer Wizard.

2. Click the dropdown arrow and click parcel01mrg.

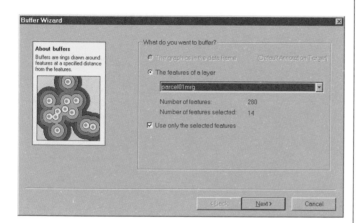

Since there are currently selected features in the layer, the wizard automatically checks the option to Use only the selected features and shows you that there are 14 selected parcels.

3. Click Next.

As with the parks, you'll buffer the residential parcels to 150 meters.

4. Double-click in the text box and type 150, then click Next.

5. On the next screen, click the options to dissolve barriers between buffers and to create buffers only outside the polygon(s).

6. Click the option to create the buffers in a new layer, make sure the path to the Analysis folder is specified, and type res01buf as the layer name. Then click Finish.

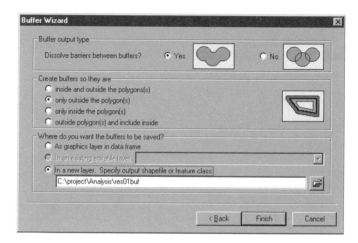

The residential parcel buffers are displayed.

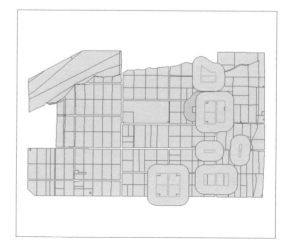

7. Click the Selection menu and click Clear Selected Features to unselect the residential parcels.

8. Uncheck the parcel01mrg and parks_polygon layers since you don't need them displayed right now.

Overlay the park and residential buffers

Now you'll combine the park and residential buffers to delineate the areas within 150 meters of a park or residence.

1. Click the Tools menu and click GeoProcessing Wizard.

2. Click the option to Union two layers.

   Union creates a layer containing the areas covered by either of the input layers.

3. Click Next.

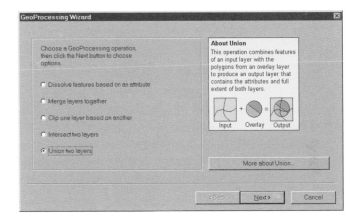

4. Click the dropdown arrows for each text box and click park02buf as the layer to union and res01buf as the overlay layer.

5. Make sure the path to the Analysis folder is specified and type respark_buf as the output layer name.

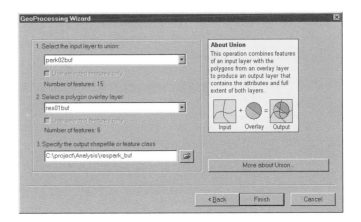

6. Click Finish.

Now the park and residential parcel buffers are combined on one layer. Next you'll combine these with the flood zone to delineate all the areas the wastewater treatment plant should be outside of.

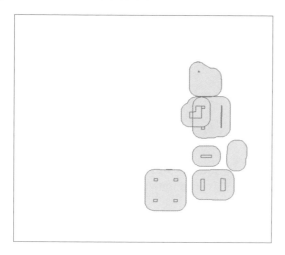

Overlay the residential/park buffers and flood zone

1. Click and drag the flood_zone layer, located in the City_layers folder, from the Catalog tree onto the map.

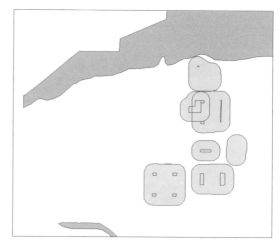

2. Click the Tools menu and click GeoProcessing Wizard.

You'll use Union again since you want to create a layer of areas inside either the residential and park buffers, or the flood zone, or both.

3. Click the option to Union two layers, then click Next.

4. Click the dropdown arrows for each text box and click respark_buf as the layer to union and flood_zone as the overlay layer.

5. Make sure the path to the Analysis folder is displayed and type respark_flood as the output layer name.

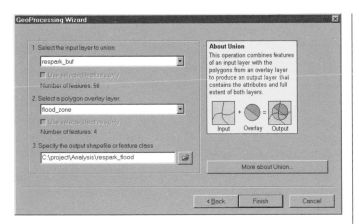

6. Click Finish.

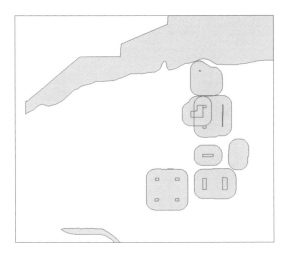

Now all the areas the plant should be outside of are combined on a single layer.

So far you've performed a series of buffers and overlays to create the two layers delineating the acceptable and unacceptable areas for the wastewater treatment plant site based on the City's criteria. You can see that even a fairly simple GIS analysis consists of stringing together a set of individual operations, often with the same operation repeated on different datasets. The operations build on the previous ones to achieve a resulting layer or layers. In the process, interim layers are created. You'll want to keep some of these layers to help verify the final results of your analysis. Others you can remove from the map.

7. Click respark_buf in the table of contents to select it, press the Ctrl key, and click res01buf and park02buf so all three layers are selected.

8. Right-click one of the selected layers and click Remove.

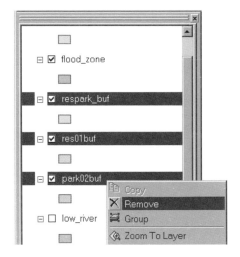

Before continuing, save your map.

9. Click File and click Save.

In the next two sections, you'll use the interim layers (low_river and respark_flood) in a series of selections to eliminate the unsuitable parcels, creating a final layer of suitable parcels. Here's the flowchart for this process:

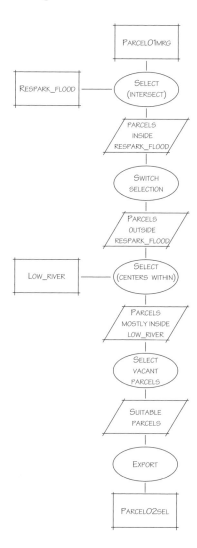

Finding the parcels that meet the location criteria

At this point you have two layers to use to select parcels meeting the criteria for the wastewater treatment plant site. First you'll select parcels falling outside the respark_flood polygons, then you'll select the subset of these that fall inside the low_river polygon.

Select the parcels outside the residential/park buffer and flood zone areas

You'll use Select By Location to select parcels that intersect the respark_flood layer. The selected parcels will lie fully or partially within the flood zone or within a residential or park buffer. You'll then switch the selected set to select the parcels that lie outside those areas. The selected parcels will be outside the flood zone and more than 150 meters from a park or residence.

1. Click the check box next to the parcel01mrg layer to display it.

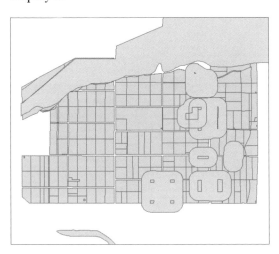

You can see that some parcels fall within the respark_flood area.

2. Click Selection and click Select By Location.

The Select By Location dialog box appears. This dialog box lets you compose a wide variety of queries to select features in one layer based on their location relative to features in another layer. In the top box you specify the type of selection. The default is to create a new selected set (select features from), which is what you want. In the next box you choose the layer to select from.

3. Scroll down and click the check box next to parcel01mrg.

Next you specify the relationship between the layers. The default is intersect—features that fall fully or partially within the features of the selection layer will be selected. This is the option you want, so accept the default. Finally, you specify the selection layer.

4. Click the dropdown arrow and click respark_flood.

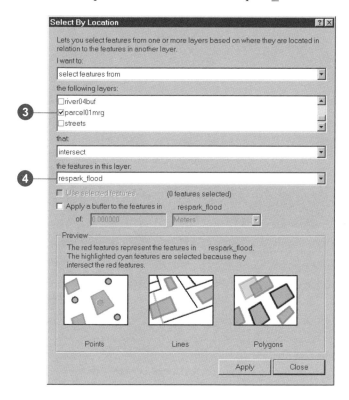

5. Click Apply at the bottom of the dialog box, then click Close to close the Select By Location dialog box.

   ArcMap selects the parcels that are completely or partially within the respark_flood polygons and highlights them on the map.

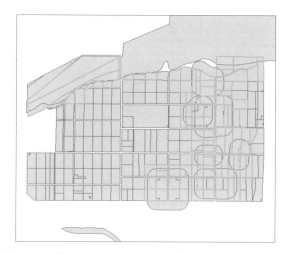

You actually want the parcels that are outside the respark_flood polygons, not inside, so you'll switch the selected set of parcels.

6. Right-click parcel01mrg in the table of contents, point to Selection, and click Switch Selection.

Now the parcels outside the flood zone and more than 150 meters from a park or residence are selected.

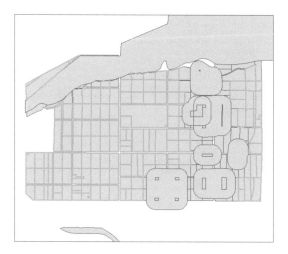

Select those parcels within the combined river buffer and lowland area

The next step is to select from the currently selected parcels those that are within the lowland area and within 1,000 meters of the river. You'll use Select By Location again, this time to select from the currently selected set of parcels.

1. Click the check box next to the low_river layer to display it.

2. Click Selection and click Select By Location.

3. Click the dropdown arrow for the top box and click "select from the currently selected features in".

4. Check the box to select features from parcel01mrg (if it's not already checked).

5. Click the dropdown arrow to choose a relationship type and click "have their center in".

   That will select the parcels that have at least half their area within the low_river polygon.

6. Click the dropdown arrow, scroll down, and click low_river as the selection layer.

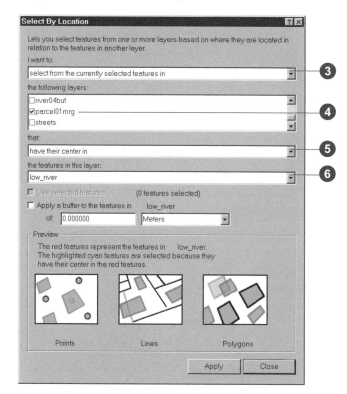

7. Click Apply and click Close to close the Select By Location dialog box.

ArcMap selects those parcels that are mostly within the low_river polygon. You can see that the selected parcels are outside the respark_flood area and inside the low_river area.

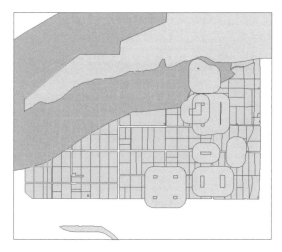

So far you've narrowed down the set of possibly suitable parcels from all the parcels in the study area to those outside the flood zone and more than 150 meters from a park or residence. You then narrowed the set further to those parcels with at least half their area inside the lowland area (elevation 365 meters or less) and within 1,000 meters of the river. The next step is to find the vacant parcels from among these.

Finding the vacant parcels

To find which parcels meet all the City's requirements as a location for the wastewater treatment plant, you'll select the vacant parcels from among the currently selected ones.

Select vacant parcels using the land use code

In the last two selections, you selected parcels based on their location. This time, you'll select based on an attribute, specifically, the parcel land use code. As you recall from the metadata, in the Assessor's database vacant land is coded with values in the 700s. You'll create a query expression to select the parcels that have a land use code greater than or equal to 700 and less than or equal to 799.

1. Click Selection and click Select By Attributes.

The Select By Attributes dialog box appears.

2. Click the dropdown arrow in the top box and click parcel01mrg as the layer to select from.

3. Click the dropdown arrow in the next box and click "Select from current selection" as the procedure to use.

Now you'll create the query expression.

4. Double-click LANDUSE in the Fields list.

5. Click the greater than or equal to (>=) sign and type "700".

6. Click And.

7. Double-click LANDUSE in the Fields list.

8. Click the less than or equal to (<=) sign and type "799".

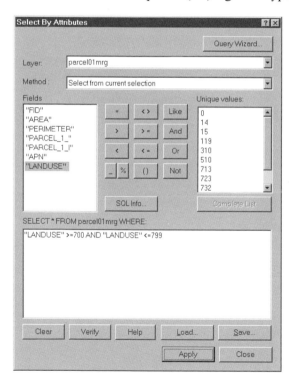

Your query expression should look like this:

"LANDUSE" >= 700 AND "LANDUSE" <= 799

9. Click Apply.

ArcMap selects the parcels that have land use codes in the 700s (the vacant parcels) and highlights them.

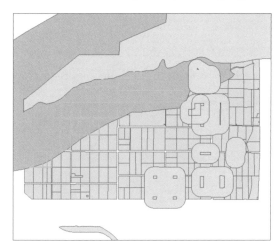

10. Close the Select By Attributes dialog box.

The selected set of parcels contains only the ones that meet the City's required criteria:

• Outside the flood zone

• At least 150 meters from parks and residences

• Elevation of 365 meters or less

• Within 1,000 meters of the river

• Vacant

Export the selected parcels to a new shapefile

To make it easier to work with only the suitable parcels, you'll export the selected parcels to a new shapefile.

1. Right-click parcel01mrg in the table of contents, point to Data, and click Export Data.

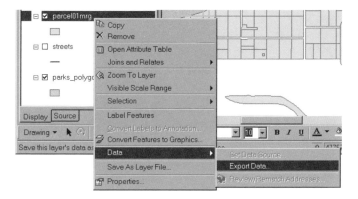

The Export Data dialog box appears. Since there are features currently selected in the parcel01mrg layer, the dialog box defaults to Selected features.

2. Make sure the path to the Analysis folder is displayed in the output feature class box.

The dialog box defaults the name of the new shapefile to be Export_Output.shp.

3. Highlight the text and type "parcel02sel" to rename the feature class.

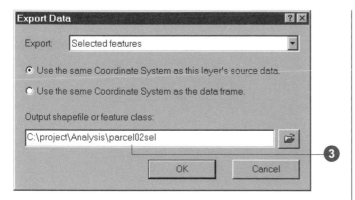

4. Click OK and click Yes when prompted to add the exported data to the map.

The new layer contains only the suitable parcels.

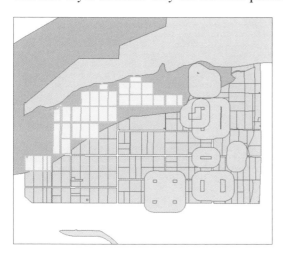

5. Click Selection and click Clear Selected Features to unselect the parcels in the parcel01mrg layer.

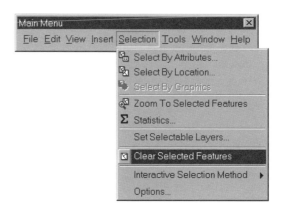

6. Click File and click Save.

Finding suitable parcels near roads and near the wastewater junction

To make a final decision on the location of the wastewater treatment plant, the City Council wants to know which suitable parcels are within 50 meters of a road and within 500 and 1,000 meters of the main wastewater junction. These will be considered the highly suitable parcels.

You'll select the parcels near these features and tag them with a code. That way you can display them color coded on your final map.

Here are the steps you'll perform:

1. Add two fields to the parcel02sel attribute table to hold the distance values: ROAD_DIST and JUNC_DIST.

2. Assign distance from roads.

• Select the parcels within 50 meters of a road.

• Assign a value 50 to the ROAD_DIST field for the selected parcels in the parcel02sel attribute table.

3. Assign distance from the wastewater junction.

• Buffer the junction to 500 and 1,000 meters.

• Select the 1,000-meter buffer and use it to select the parcels within 1,000 meters of the junction.

• Assign a value of 1,000 to the JUNC_DIST field for the selected parcels in the parcel02sel attribute table.

• Select the 500-meter buffer and use it to select the parcels that fall within 500 meters of the junction.

• Assign a value of 500 to the JUNC_DIST field for the selected parcels.

Here's the flowchart of the process:

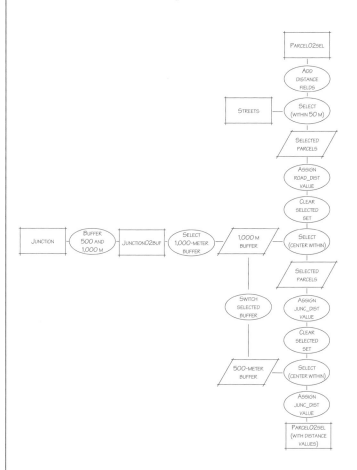

Add fields to the parcels layer

Before you find the parcels near roads and the wastewater junction, you'll add two fields to the parcel02sel attribute table to contain the distance values you'll assign.

1. Right-click parcel02sel and click Open Attribute Table.

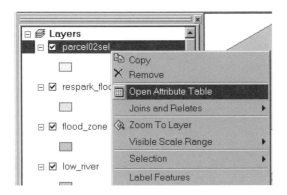

2. Click the Options button on the table and click Add Field.

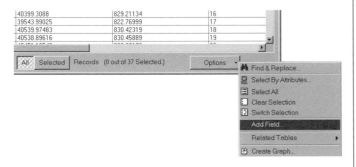

The Add Field dialog box appears.

3. Type ROAD_DIST as the field name.

You can accept the default type of short integer.

The short integer type stores values up to 32,768, which is sufficient for the ROAD_DIST field—the values in this field will be either 50 or 0.

4. Click OK.

Now add the JUNC_DIST field in the same way.

5. Click Options and click Add Field.

6. Type JUNC_DIST as the field name and click OK.

7. Scroll all the way to the right to see the new fields.

APN	LANDUSE	ROAD DIST	JUNC DIST
029204402	732	0	0
029204408	732	0	0
029204407	732	0	0
029204406	732	0	0
029204409	732	0	0
029204410	732	0	0
029204411	732	0	0
029204412	732	0	0

Currently, the two columns have no values in them, or contain zeros, since you just added the fields. In the next few steps you'll select the parcels near roads and the wastewater junction and enter values for the selected parcels.

8. Close the attribute table for now.

Find parcels within 50 meters of a road

The City would prefer that the parcel for the new plant be within 50 meters of an existing road. You'll use the streets layer to select the parcels within 50 meters of a road and assign a value of 50 to the ROAD_DIST field.

1. Before continuing, uncheck the check boxes for all layers except parcel02sel in the table of contents so that only the suitable parcels are displayed.

2. Check the streets layer to display it.

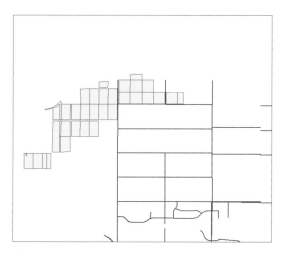

3. Click the Selection menu and click Select By Location.

You've seen this dialog box before. This time, you'll select features on one layer (parcels) within a distance of features on another layer (streets).

4. Click the dropdown arrow next to the "I want to" box and click "select features from".

5. Click the check box for parcel02sel.

6. Click the dropdown arrows for the next two text boxes and click "are within a distance of" and "streets".

The option to "Apply a buffer to the features in streets" is automatically checked.

7. Type "50" in the text box to select parcels using a buffer of 50 meters.

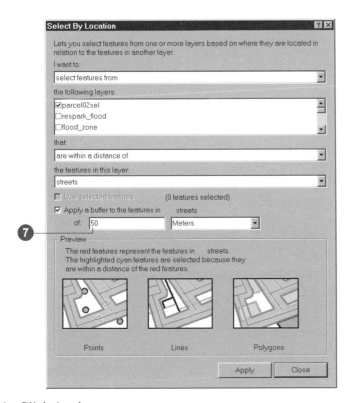

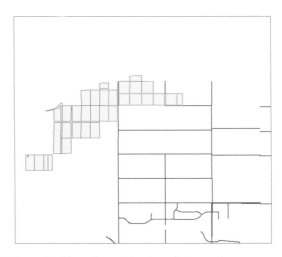

Using a buffer when selecting features is an easy way of finding features within a certain distance of other features. This method is quicker than using the Buffer Wizard (as you did with rivers, parks, and residential parcels) when you don't need to create a separate buffer layer to combine with other layers.

9. Click Close to close the Select By Location dialog box.

Now you can assign a value to the ROAD_DIST field for the selected parcels to tag them as being within 50 meters of a road.

Assign a value to the ROAD_DIST field

To assign or update values in a layer's attribute table, you need to open the layer for editing. You can edit the attributes either by using the Attributes button on the Editor toolbar, as you did for the historic park, or by editing directly in the attribute table, as you'll do here. To assign values in the attribute table, you create a calculation expression. The values are assigned to the selected parcels or to all the parcels if none are selected.

8. Click Apply.

The parcels within 50 meters of streets are selected.

1. On the Editor toolbar, click the Editor dropdown arrow and click Start Editing (click the Editor Toolbar button, if necessary, to open the toolbar).

2. Click the Analysis folder as the folder to edit data from and click OK.

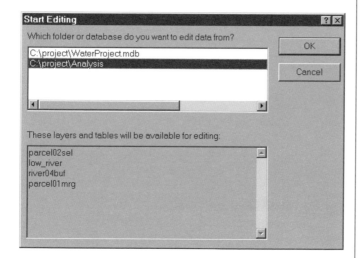

3. Click the Target dropdown arrow and click parcel02sel as the layer to edit.

4. Right-click parcel02sel in the table of contents and click Open Attribute Table.

   The selected parcels (those within 50 meters of a road) are highlighted.

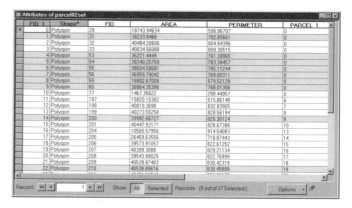

5. Scroll to the right in the Attributes window, right-click the ROAD_DIST field (the cursor changes to a down arrow when it's over the field name), and click Calculate Values.

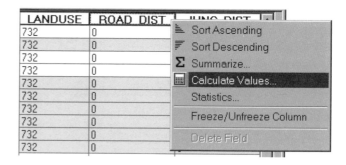

The Field Calculator dialog box appears. Since you clicked the ROAD_DIST field, ArcMap starts the calculation expression for you by displaying "ROAD_DIST =".

6. Click in the expression box and type "50" to complete the expression.

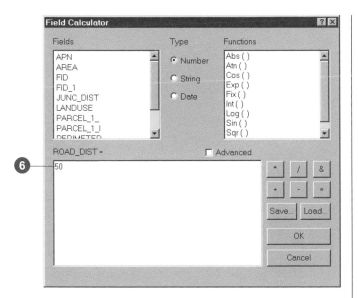

7. Click OK.

ArcMap assigns a value of 50 to the ROAD_DIST field for the selected parcels—the ones within 50 meters of a road. All other parcels have a value of 0 for ROAD_DIST. You'll use these values along with the JUNC_DIST values to color code the highly suitable parcels on your final map.

APN	LANDUSE	ROAD_DIST	JUNC_DIST
029204402	732	0	0
029204408	732	50	0
029204407	732	0	0
029204406	732	0	0
029204409	732	50	0
029204410	732	50	0
029204411	732	50	0
029204412	732	50	0
029204118	732	50	0
029204119	732	50	0

8. Click the Editor dropdown arrow on the Editor toolbar and click Stop Editing. Click Yes when prompted to save your edits.

When you save your edits, ArcMap clears the selected set of features, so you'll be able to start with the full set of suitable parcels to find the ones near the wastewater junction in the next step.

Keep the parcel02sel attribute table open since you'll need it in the next step, but you may want to move, resize, or minimize it so you can see the map.

Measure and assign the wastewater junction distance

The City would prefer that the plant site be within 1,000 meters of the point where the plant will connect to the existing wastewater system. The council is willing to accept parcels that extend beyond the 1,000-meter buffer as long as most of the parcel is within the buffer. Parcels within 500 meters are even more desirable.

You need to find the parcels within 500 meters and 1,000 meters of the main wastewater junction and tag them with the distance. For this task you'll use a number of the tools you used previously: buffer, selection, and attribute editing. Rather than give you step-by-step instructions, we've listed the major steps to give you a chance to work through the task yourself. You can review the previous sections if you need help with the specific steps.

Add the wastewater junction coverage to the map (it's located in the utility folder under the City_share folder). The junction coverage contains the main wastewater junction point where the new plant will connect to the existing system.

Use the Buffer Wizard to buffer the junction to 500 meters and 1,000 meters. You'll specify that the buffer be created as multiple rings. There will be two rings, and the distance between them is 500 meters. You can use the option not to dissolve between the buffers (since there will be no overlap, anyway). Create a new layer in the Analysis folder and name it junction02buf.

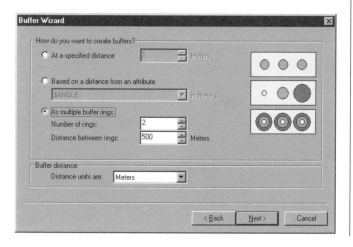

The buffers draw on the map but obscure the junction and the parcels. Change the symbology of the junction02buf layer so the buffers are drawn with no shading (use "no color" as the fill color).

Now you can see which parcels are within 0 to 500 meters and within 500 to 1,000 meters of the junction. Next you'll select each set of parcels and tag them with the respective distance.

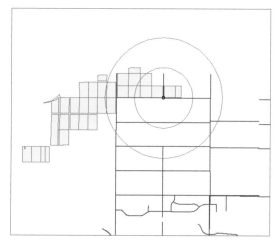

Use the Select Features tool to select the 500- to 1,000-meter buffer by pointing at it. Then use Select By Location to select the parcels in parcel02sel that have their center in the features of junction02buf. This will select all the parcels between 500 and 1,000 meters from the junction. Close the Selection dialog box when you're done.

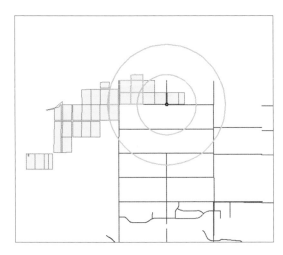

Now switch the selected features for junction02buf so the 0- to 500-meter buffer is selected (or simply use the Select Features tool to select the inner buffer). Then select by location to select the parcels that have their center in the selected features of junction02buf (i.e., the 0- to 500-meter buffer). Now you'll have only those parcels within 500 meters of the junction.

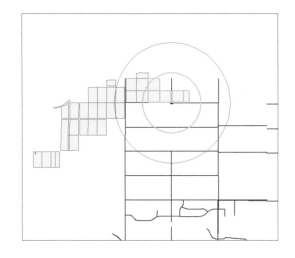

Use the Editor toolbar to start editing parcel02sel (make sure parcel02sel is the target layer). Open the attribute table for parcel02sel (click the Restore button if you minimized it before) and assign the value 1000 to the JUNC_DIST field for the selected parcels. Save your edits, but don't stop editing yet. You'll notice that some of the suitable parcels are both within 50 meters of a road and 1,000 meters of the junction, a few are more than 50 meters from a road but within 1,000 meters of the junction, and quite a few are not near either (both fields have values of 0).

APN	LANDUSE	ROAD DIST	JUNC DIST
029204402	732	0	1000
029204408	732	50	1000
029204407	732	0	1000
029204406	732	0	0
029204409	732	50	1000
029204410	732	50	1000
029204411	732	50	0
029204412	732	50	0
029204118	732	50	0
029204119	732	50	0

Assign the value 500 to the JUNC_DIST field for the selected parcels.

APN	LANDUSE	ROAD_DIST	JUNC_DIST
029204402	732	0	1000
029204408	732	50	1000
029204407	732	0	1000
029204406	732	0	500
029204409	732	50	1000
029204410	732	50	1000
029204411	732	50	500
029204412	732	50	500
029204118	732	50	500
029204119	732	50	500

Stop editing and save your edits. Then save your map.

Now you've assigned the distance from the wastewater junction to those parcels within 500 and 1,000 meters. On your final map, you'll color code the parcels based on their distance to roads and the wastewater junction so the City Council and the public can see which are the highly suitable parcels.

Another way of selecting the parcels near the junction would have been to use select by location with a buffer distance, but by creating the junction02buf layer you'll be able to display the rings on the map to make it easier for the City Council and public to visualize the distance from the wastewater junction.

Finding suitable parcels meeting the required total area

The final step in the analysis is to find suitable parcels that are large enough to construct the wastewater treatment plant on. The minimum required area for constructing the plant is 150,000 square meters. You'll check the parcel02sel attribute table to see which parcels are at least this large.

Sort the parcels by AREA

The parcel02sel attribute table should still be displayed.

1. Scroll to the left, if necessary, to see the AREA field.

2. Right-click AREA and click Sort Descending.

AREA		PERIMETER
18743.94634	≡ Sort Ascending	
38233.8468	⊏ Sort Descending	
40484.20606	Σ Summarize...	
40834.56088	▦ Calculate Values...	
36221.4449		
38340.25769	Statistics...	
38504.58681		
36959.79042	Freeze/Unfreeze Column	
18802.67008		
36864.35396	Delete Field	
1467.35622	288.44957	

The largest parcels are listed at the top.

AREA	PERIMETER
61280.49249	1274.63154
41476.83105	837.99145
41162.43451	839.65045
40834.56088	808.30515
40819.3698	832.83905
40766.12503	831.8554
40539.97483	830.42319
40538.89616	830.45889

None of the suitable parcels is anywhere near 150,000 square meters in size. In fact, the largest one is only a little over 60,000 square meters (you may need to scroll to the top of the table to see it). It looks like the City is going to have to assemble the plant site from several parcels or else relax their criteria to include more parcels as possible sites. You'll check to see if there are several adjacent suitable parcels that total 150,000 meters.

Check for adjacent parcels totaling 150,000 sq. m

You'll first identify several parcels to see how large they are, then select a group of them to see if they total 150,000 square meters in area.

1. Right-click parcel02sel in the table of contents and click Zoom To Layer, then uncheck junction02buf so it's no longer displayed. Move the attribute table to better see the parcels, if necessary.

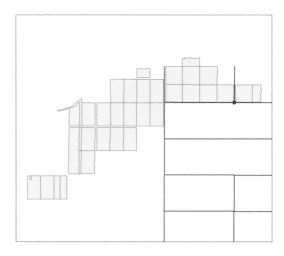

2. Click View and click Identify Results.

3. Click the Layers dropdown arrow in the Identify Results box and click parcel02sel.

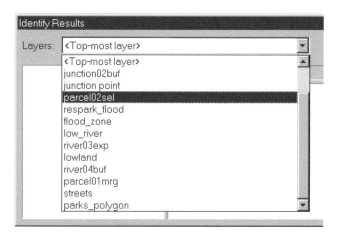

That sets parcel02sel as the layer to identify features in. Leave the Identify Results box open.

4. Click the Identify tool.

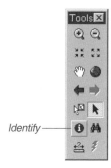

5. Click the parcel adjacent and just to the west of the wastewater junction.

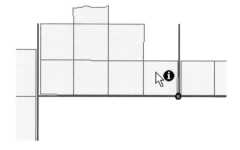

You can see that this parcel is just under 37,000 square meters in area.

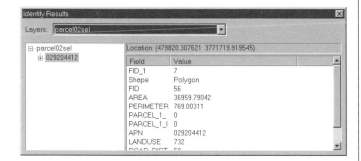

6. Click the parcel to the left of the previous one.

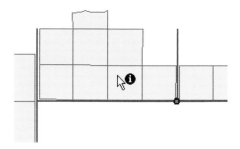

This parcel is about 38,500 square meters. You can see that most of the suitable parcels near the junction are about the same size. It looks like four contiguous parcels together will total about 150,000 square meters. Close the Identify Results box.

7. Click the Select Features button.

8. Click and drag a box around the intersection of four contiguous parcels.

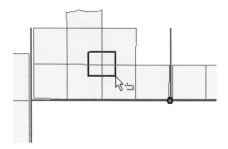

The parcels are highlighted on the map and in the attribute table.

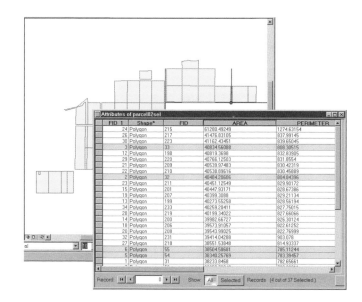

9. Right-click AREA in the table and click Statistics.

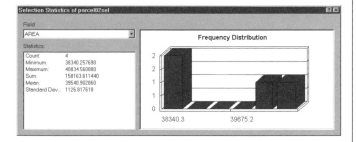

AREA		PERIMETER
61280.49249	≜ Sort Ascending	
41476.83105	≡ Sort Descending	
41162.43451		
40834.56088	Σ Summarize...	
40819.3698	▦ Calculate Values...	
40766.12503	Statistics...	
40539.97483		
40538.89616	Freeze/Unfreeze Column	
40484.20606		
40451.12549	Delete Field	

The Selection Statistics dialog box appears.

ArcMap calculates summary statistics about the selected parcels and presents a graph of the distribution of values. You're not interested in the distribution of values in this case, but the statistics are useful. You can see the number of parcels selected (count), the size of the smallest and largest parcels, the total area, and the mean size of the four parcels.

The sum of the areas of these four parcels is just over 158,000 square meters. There is enough room for the plant here. Further examination shows that there are a number of combinations of adjacent suitable parcels that will provide enough room for the plant. (You can use the Select Features tool to draw a box around different groups of parcels or select a parcel by clicking on it and then clicking additional parcels while pressing the Shift key to add them to the selected set. Then recalculate the statistics to see the total area.)

10. Close the Selection Statistics dialog box and the attribute table when you're done selecting parcels.

11. Click Selection and click Clear Selected Features.

So it looks like the City will be able to piece together a site for the plant. You decide to set up an interactive session at the City Council meeting using a big screen so you can query different combinations of parcels on request from the council members.

Selection Statistics of parcel02sel

Field: AREA

Statistics:
Count:	4
Minimum:	38340.257690
Maximum:	40834.560880
Sum:	158163.611440
Mean:	39540.902860
Standard Dev...	1126.817618

Frequency Distribution

Reviewing the analysis results

Anticipating one question the City Council is likely to ask, you decide to see if there are any parcels in the study area at least 150,000 sq. meters in size and, if so, why they didn't make the final cut.

You'll use select by attribute to find any parcels greater than 150,000 square meters, then display them with the analysis layers to see which criteria they do or don't meet.

Find any parcels at least 150,000 square meters in size

1. Uncheck parcel02sel and streets and check parcel01mrg so you can see all the parcels in the study area.

2. Right-click parcel01mrg in the table of contents and click Zoom To Layer.

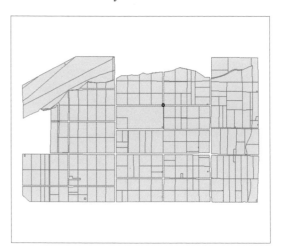

3. Click Selection and click Select By Attributes.

4. Click the Layer dropdown arrow and click parcel01mrg. Use the default method (Create a new selection).

5. Double-click AREA in the Fields list, click the greater than or equal to sign (>=), and type "150000" to create the query expression.

The expression should look like this:

"AREA" >= 150000

6. Click Apply, then click Close to close the dialog box.

There are three parcels of at least 150,000 square meters.

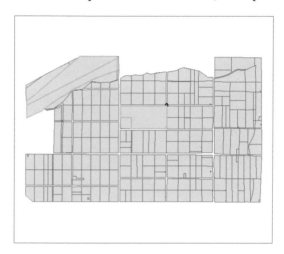

First check to see if the parcels are vacant.

7. Right-click parcel01mrg in the table of contents and click Open Attribute Table.

8. Click Selected at the bottom of the table window to show just the selected parcels.

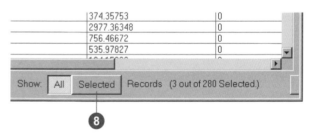

Two of the parcels have no land use code, but the other has a code of 732, which is vacant (you may need to scroll to the right to see the LANDUSE field).

PARCEL 1 1	APN	LANDUSE
0	029201137	0
0	029201142	0
0	029204208	732

9. Click the box next to the vacant parcel to highlight it. If necessary, minimize or move the table so you can see the highlighted parcel on the map.

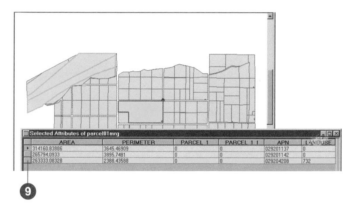

So the two parcels in the upper left of the study area were rejected because they weren't classified as vacant land. You'll have to check with the City Assessor to find out the actual land use for the parcels—it might be that they're vacant but were never assigned land use codes in the database. In the meantime, though, you'll check to see why the third parcel was rejected.

10. Close the attribute table.

Display the selected parcels with the criteria layers

1. Check the respark_flood layer to display it.

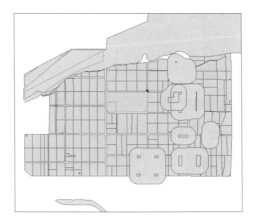

You can see that the two parcels in the upper left are mostly within this layer, which includes the flood zone and buffers around parks and residential areas.

2. Check the flood_zone layer, then click and drag it above respark_flood in the table of contents.

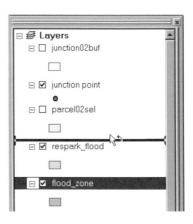

Now it's clear that the two parcels are outside the residential and park buffers but are within the flood zone. So even if these parcels are vacant, they'd be rejected for being inside the flood zone. The third parcel, though, is outside the flood zone and the park and residential buffers.

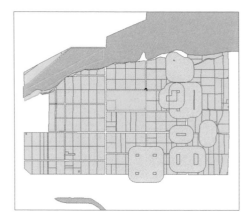

3. Uncheck respark_flood and flood_zone and check low_river.

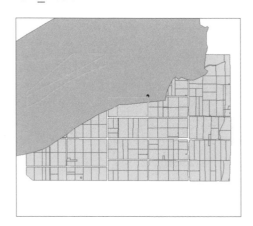

The City required that suitable parcels be completely or mostly within this area. More than half of this parcel is outside the area.

4. Check the lowland layer, then uncheck low_river.

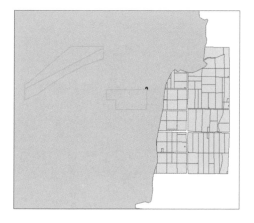

The parcel is fully within the lowland layer, so it must be the river buffer that it's mostly outside of (that is, most of the parcel is more than 1,000 meters from the river).

5. Uncheck lowland and check the river04buf layer.

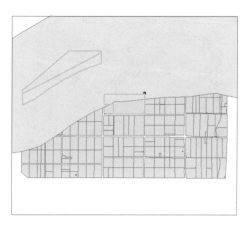

Yes—that's why it was rejected. Still, a portion of the parcel is within the buffer, and the parcel is also adjacent to the wastewater junction, a big plus. You decide to highlight it on the final map to call it to the City's attention as a possible alternate site. Any additional construction costs may be offset by the cheaper land costs involved in buying a single parcel rather than four separate ones.

Create a layer containing the alternate site

1. Click the Select Features tool.

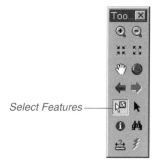

Select Features

2. Click inside the alternate site (but outside the river buffer) so it is the only parcel selected.

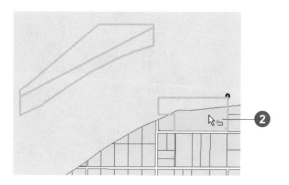

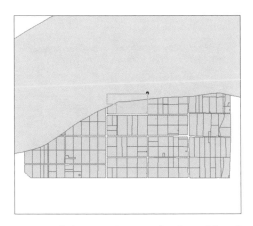

3. Right-click parcel01mrg in the table of contents, point to Selection, and click Create Layer From Selected Features.

ArcMap adds the layer containing the single parcel to the map.

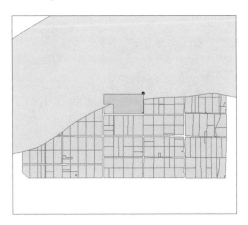

4. Click the layer name (parcel01mrg selection) to select it, then click again to highlight the name.

5. Type "alternate site" as the new layer name and press Enter.

This is a temporary layer for this map—it is not stored as a separate layer file. In the next chapter you'll change its symbology and add it to the final map.

Clean up the table of contents

You won't be using several of the analysis layers on your final map, so you can remove them at this point.

1. Click flood_zone, then press the Ctrl key and click respark_flood, low_river, lowland, river04buf, and parks_polygon so they're all selected.

2. Right-click one of the selected layers and click Remove.

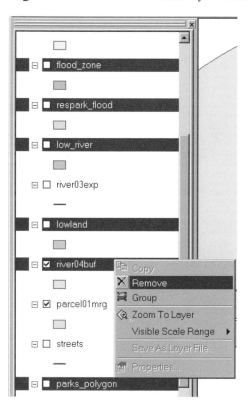

At this point, your map should only include the following layers:

alternate site

junction02buf

junction point

parcel02sel

river03exp

parcel01mrg

streets

If any other layers are still on your map, go ahead and remove them.

3. Click File and click Save to save your map.

The analysis phase of the project is finished, at least for now. GIS makes it easy to modify your criteria and rerun the analysis, if necessary.

There are many questions that GIS analysis can help you answer and many different approaches. This analysis showed one approach to solving a particular problem using several of the most common GIS analysis tools: buffer, overlay, and selection. You'll combine these tools and others in various ways when performing your own analyses.

In the next chapter you'll make a map to present the results of your analysis to the City Council and the public.

Presenting the results

8

In this chapter you'll create a poster-sized map to present the results of your analysis. The poster will contain three maps. One map will show the geographic relationship of the suitable parcels to the rest of the City. Another map will show all of the suitable parcels. The third map will show the highly suitable parcels symbolized according to their proximity to the main wastewater junction and to roads. You will label these parcels with their parcel identification numbers.

You'll also create a report showing parcel identification numbers, area, and distance from the junction for the highly suitable parcels.

The map will also include explanatory text, a North arrow, legends, scale bars, and a title.

Designing the map

You'll want to give some thought to the map design before you start laying out the map. The design should reflect how the map will be used and who the audience for the map is. In this case, your map will be displayed at a City Council meeting. The council members are probably familiar with the issues involved in siting the wastewater treatment plant, but the members of the public attending the meeting may not be. Both groups will want to see the location of the suitable parcels in relation to the rest of the City. They'll also want to see all the suitable parcels, as well as the most likely candidates for the plant site, with additional information about the highly suitable parcels.

You'll first want to decide what elements will be on the map and list them. Then decide how the elements will be arranged on the page.

In this case, you'll create three maps on one poster-size page for display at the City Council meeting.

1. An overview map of the City showing the location of the study area and including the following layers:

*   Streets (streets.lyr)
*   The river (river03exp)
*   The elevation grid (elevation_grid.lyr)
*   The study area extent (graphic rectangle)

2. A map of the study area showing all suitable parcels and including the following layers:

*   Suitable parcels in one color (parcel02sel)
*   Other parcels in a different color (parcel01mrg)
*   The alternative site shaded using diagonal hatching (alternate site)
*   The wastewater junction (junction point)
*   The 500- and 1,000-meter buffers around the junction (junction02buf)
*   The river (river03exp)

3. A map of the highly suitable parcels including the following layers:

*   Highly suitable parcels color coded by distance from roads and from the wastewater junction and labeled with their parcel numbers (parcel02sel)
*   All other suitable parcels in a neutral color (parcel02sel)
*   The alternate site shaded using diagonal hatching and labeled with its parcel number and area (alternate site)

You'll also want to include other map elements and explanatory text to make the map easier to understand and more attractive.

The additional map elements are as follows:

- Report listing the highly suitable parcels
- Text block listing the site selection criteria
- Map title
- Scale bar for each data frame
- Legend for each data frame
- North arrow
- City logo
- Map reference information
- Graphic rectangles to complete the map composition

Once you've decided on what the maps should show and what other elements to include, you'll want to decide how the elements will be arranged on the page. While it's very easy in ArcMap to move and resize the individual maps and other elements on the page, you may want to make a sketch on paper to use as a guide. The sketch should show at least the approximate position of the maps and map elements and contain notes about the page size and map sizes. You can reposition and resize the elements interactively as you create the map.

Here's a sketch for the map you'll create.

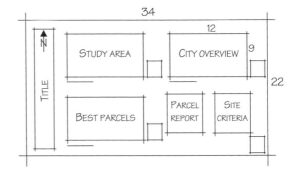

Here are the major steps you'll complete to create the map:

- Create the three data frames.
- Modify the data frames to show the required layers and geographic extent.
- Create and add the parcel report.
- Add the text block listing the site criteria.
- Add the legends and scale bars for each data frame.
- Add the other map and graphic elements (North arrow, title, logo, map reference information, graphic rectangles).

Setting up the map page

You'll be creating a poster-sized map with three data frames, one for each map. You'll work in both data view and layout view to create the map.

If necessary, start ArcMap and open the water project.mxd map (located in the project folder). The map should currently have the alternate site, junction point, and parcel01mrg layers displayed.

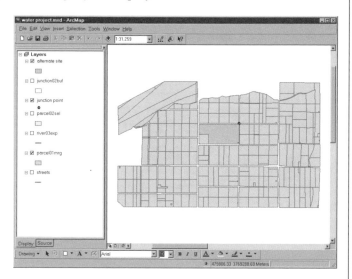

Switch to layout view

First you'll switch from data view to layout view.

1. Click View and click Layout View.

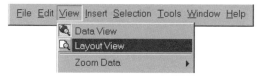

The map switches to layout view and displays the map page filled by a data frame containing the currently displayed layers. The Layout toolbar also appears.

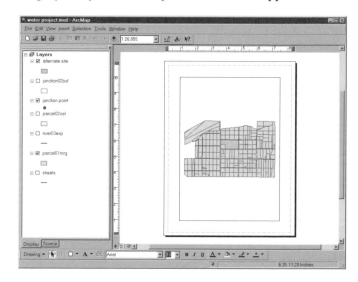

Layout view lets you view several data frames on a single page and interactively work with the map elements. A data frame is a way of organizing layers together on a map page. Currently there is only one data frame on the page, indicated by the rectangle.

The Layout toolbar contains tools for zooming and panning on the page itself. The tools on the Tools toolbar let you work with the data inside the data frame, just as you do in data view.

Zoom in on map page

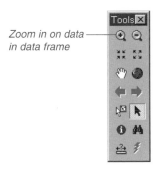

Zoom in on data in data frame

Change the page size

You'll set the page size to accommodate the final printed map.

1. Right-click on the page, outside the data frame, and click Page Setup (if you right-click inside the data frame, you open the data frame Properties dialog box).

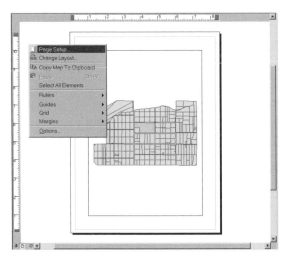

The Page Setup dialog box appears. The Same as Printer box is checked, indicating that ArcMap will automatically detect the printer's page size.

The City has another printer that is capable of printing D size (22 x 34 inch) pages, so you don't want the page size to be the same as the default printer.

2. Uncheck Same as Printer.

3. Click the Standard Page Sizes dropdown arrow and click D.

Now the virtual page for the map will be 22x 34 inches—D size.

You want the map poster to be wider than it is high, so change the page orientation.

4. Click Landscape, then click OK.

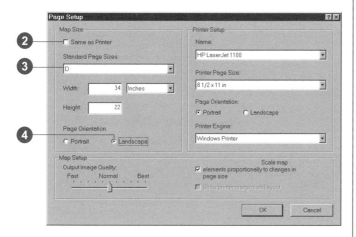

ArcMap adjusts the size and orientation of the page.

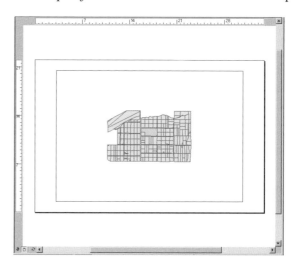

You'll use the existing data frame to display the suitable parcels. In the next steps you'll resize the data frame and make a copy of it to display the City overview map.

Resize the data frame

First you will make this data frame smaller.

1. Click the Select Elements button.

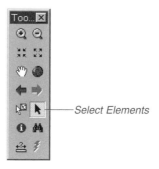

Select Elements

2. Right-click near the middle of the data frame and click Properties.

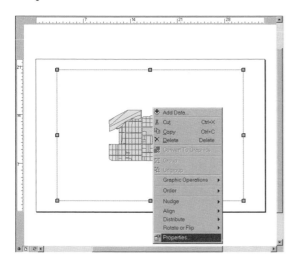

The frame is highlighted, selection handles appear at the corners and sides of the frame, and the Data Frame Properties dialog box appears.

3. Click the Size and Position tab.

You can move and resize elements on the page by clicking and dragging the whole element or one of its selection handles, or you can set the position and size by typing the values in page units in the Data Frame Properties dialog box. You'll set the size and position of the first data frame explicitly by typing the values.

You want the upper-left corner of the data frame to remain where it is, but resize the data frame to be 12-by-9 inches.

4. Click the upper-left anchor point in the Position panel.

5. In the Size panel, type 12 in the Width box and 9 in the Height box. Click OK.

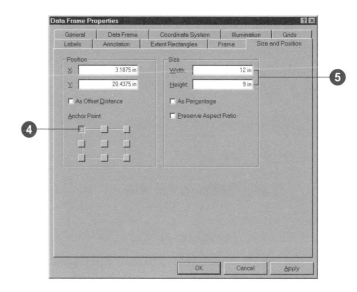

The data frame is resized, and the layers are scaled down to fit inside.

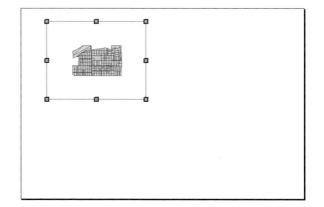

Copy the data frame

Now you'll make a copy of the data frame to contain the City overview map.

1. Click Edit and click Copy.

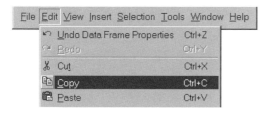

2. Click Edit and click Paste.

   The copy of the data frame is pasted onto the map on top of the original data frame.

3. Click on the data frame and drag the copy to the right of the original.

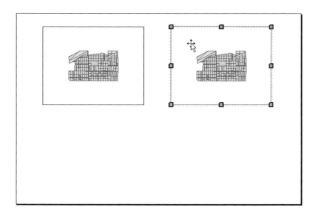

Both data frames display the same layers.

You'll use the new data frame to show the position of the project area with respect to the rest of the City and the original one to show all of the suitable parcels. You'll add a third data frame to show the highly suitable parcels, but first rename the other two.

Rename the selected data frame

The table of contents is divided into sections by data frame. Both data frames in the table of contents have the same title—"Layers" (the default)—because you copied one to create the other.

The data frame that you have just pasted onto the map is still selected—you can see its selection handles in the layout view.

1. Scroll through the table of contents until you see the Layers heading in bold type.

This is the data frame you just added. The bold type identifies this data frame as the selected data frame.

2. Click the bold Layers heading in the table of contents to select it.

3. Click Layers again to highlight the name.

(If you click twice rapidly—a "double-click"—on the name of a data frame, you will open the Data Frame Properties dialog box. You do not need to change any properties of the data frame at the moment, so if you get this dialog box, just click Cancel and try again.)

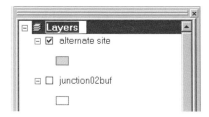

You can now type a new name for the data frame.

4. Type "City Overview" and press Enter.

Rename the original data frame

Now you'll rename the other data frame.

1. Scroll through the ArcMap table of contents until you see the Layers heading.

2. Click Layers to select the data frame, then click it again to highlight the name.

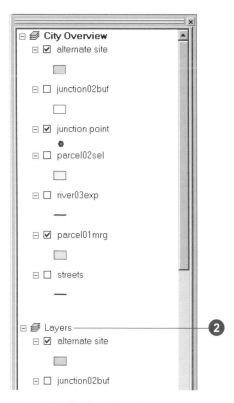

3. Type "Study Area" and press Enter.

Insert a new data frame

Now add the third data frame, which will show the highly suitable parcels. You'll do this by inserting a new data frame rather than copying an existing one.

1. Click Insert and click Data Frame.

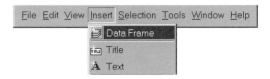

The new data frame appears as the selected data frame in the center of the map and is listed at the bottom of the table of contents, named New Data Frame 2.

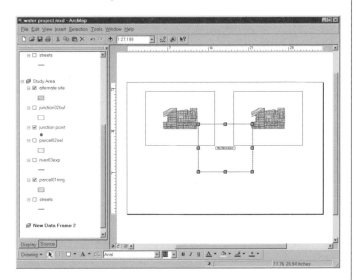

2. Click New Data Frame 2 in the table of contents to select it, then click again to highlight the name.

3. Rename the data frame by typing Best Parcels, then press Enter.

You want the new data frame to be the same size as the others (12-by-9 inches) and positioned below the Suitable Parcels data frame, so you'll resize and move it.

4. Right-click inside the new data frame on the map and click Properties.

5. Click the Size and Position tab.

6. Double-click in the Width text box and type 12, then double-click in the Height text box and type 9. Click OK.

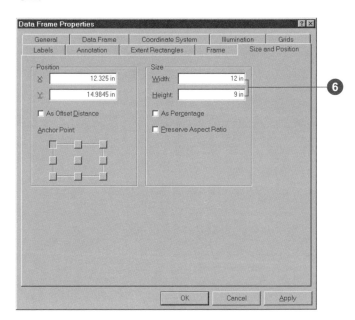

The data frame is now the same size as the others.

7. Click and drag the data frame so it is positioned below the Study Area data frame. (It's OK if it's not lined up exactly—you'll align the data frames later.)

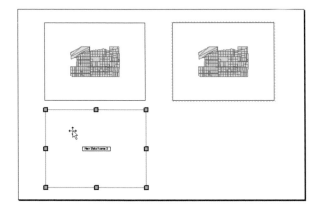

The data frame is currently empty. You'll copy the layers you need from the Study Area data frame later, after you've changed their symbology.

8. Click File and click Save to save your map so far.

Through the rest of the chapter, you'll want to periodically save your map in case you need to stop or take a break and come back to it. We'll remind you at the end of each section, but you may want to save more often.

At this point, you've added the three data frames to create the basic structure of the map. Now you'll modify each data frame to display the necessary layers. In the next step you'll change the contents of the City Overview data frame to show the location of the suitable parcels with respect to the rest of Greenvalley.

Creating the overview map

You want the City Overview data frame to show where the suitable parcels are relative to the rest of Greenvalley. Because most Greenvalley residents are familiar with the major streets of Greenvalley, you can use the streets to orient the map readers. You'll also display the river and the elevation grid so people can see that the study area containing the parcels is near the river and in the low-lying area. Later you'll add a rectangle showing the location of the study area.

To create the map, you'll remove the unneeded layers from the City Overview data frame, change the extent of the area displayed in the data frame, and change the way the streets are displayed. Then you'll change the way the river is displayed and, finally, add the elevation grid layer underneath.

Remove unneeded layers from the data frame

It is often easier to work with the data in a given data frame in data view, particularly when the layout contains several data frames.

1. Click inside the City Overview data frame on the map page to select it.

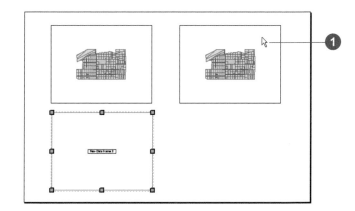

2. Click View and click Data View.

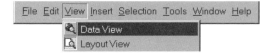

On a map with multiple data frames, switching to data view shows you the data for the currently selected data frame. In this case, the selected data frame is City Overview.

3. Scroll to the City Overview data frame in the table of contents, if necessary.

4. Click alternate site under the data frame heading in the table of contents to select it.

5. Press the Ctrl key and click the following layers to select them as well:

junction02buf

junction point

parcel02sel

parcel01mrg

6. With the layers selected, point to one of the highlighted layers, right-click, and click Remove.

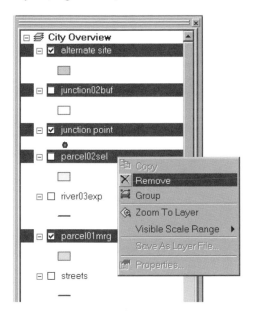

The layers are removed from the map. You could have left the layers in the data frame and just not displayed them, but it's easier to work with the data frame without the layers in the table of contents.

You should now have only streets and river03exp in the data frame. Neither are currently displayed. If any additional layers remain, remove them as well.

7. Click the check boxes next to river03exp and streets to display them.

8. Click the Full Extent button on the Tools toolbar.

This zooms to the full extent of the remaining layers.

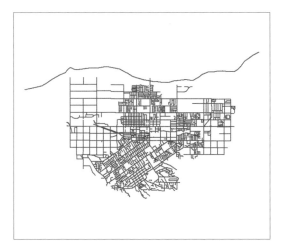

Show the major streets

The streets layer shows all of the streets of Greenvalley. Displaying only the major streets will be sufficient for showing the location of the suitable parcels and, in fact, will make the map easier to read. You'll modify the properties of this layer to simplify the representation of the streets.

1. Double-click streets under the City Overview data frame in the table of contents.

Double-clicking a layer is a quick way to get to its Layer Properties dialog box.

2. Click the Definition Query tab.

3. Click the Query Builder button.

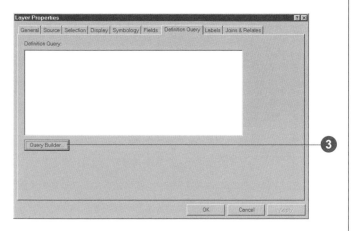

The Query Builder dialog box appears. It's similar to the other query dialog boxes you've already seen.

The Greenvalley streets in this database belong to three classes. Classes 3 and 4 are major streets; class 5 streets are smaller streets. You will select the major streets.

4. Double-click [Type].

5. Click the less than or equal to button (<=).

6. Double-click 4, review the query expression, and then click OK.

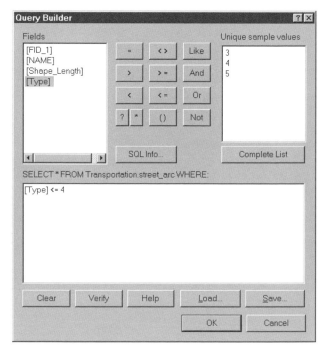

The query expression is added to the Definition Query tab of the Layer Properties dialog box. It should look like this:

[TYPE] <= 4

7. Click OK on the Layer Properties dialog box.

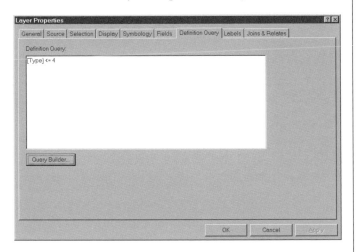

Only the major streets are displayed—the smaller streets are left off of the map. Using Definition Query is a quick way of showing only certain features in a layer without selecting them and creating a separate layer.

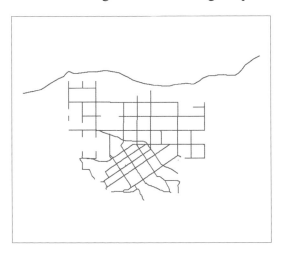

Now you can zoom to the area covered by the major streets.

8. Right-click streets in the table of contents and click Zoom To Layer.

Change the street symbol

Because you have not yet specified a symbol for the streets, they are drawn in a randomly selected color. You want the streets to be drawn with a simple black line.

1. In the table of contents, click the line symbol below the streets layer (you may need to scroll through the table of contents to locate it).

Note that streets are also in the Study Area data frame—make sure you're working with the layers under the City Overview data frame.

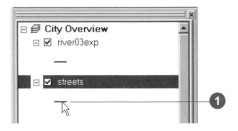

2. Click the Major Road symbol and click OK.

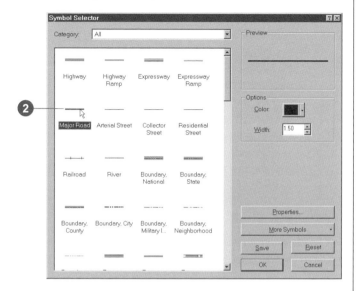

The major streets of Greenvalley are now drawn on the map with a black line.

Display the river and elevation layers

You also want to display the river and elevation grid so that the City Council and the public can see that the study area was chosen based on its proximity to the river and its location in a low-lying area.

1. Click the line symbol beneath the river03exp layer name.

2. Click the River symbol and click OK.

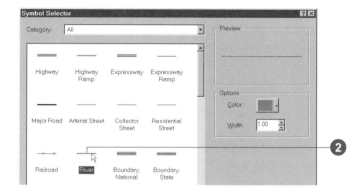

The river is now drawn with a blue line.

3. Click the Add Data button, navigate to the City_layers folder, and click elevation_grid.lyr. Click Add.

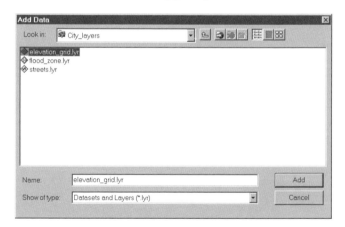

If you get a warning message about the coordinate system for the layer, click OK.

The elevation grid is added to the data frame and is displayed using the colors you specified when you created the layer.

The City Overview should now display the following layers, listed in this order in the table of contents:

river03exp

streets

elevation

4. Click View and click Layout View.

The City Overview data frame on the map now shows the major streets, the river, and the elevation grid. ArcMap attempts to center the layers in the data frame. Since the river is along the top edge, it is only partially visible.

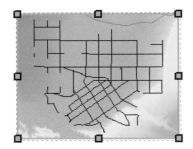

5. Click the Pan tool on the Tools toolbar.

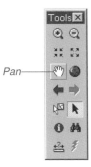

6. Click in the data frame and drag the layers down so the river is visible near the top of the data frame.

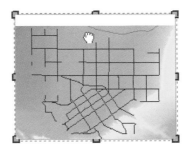

The data frame is finished for now.

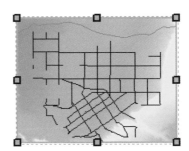

Later in the chapter you'll add a rectangle to this frame to show the location of the study area.

7. Click File and click Save to save your map.

In the next step, you'll make the necessary changes to the Study Area data frame to show the suitable parcels.

Creating the map of suitable parcels

This map will show suitable parcels shaded in one color and all other parcels in another color. It will also show the location of the wastewater junction, along with the 500- and 1,000-meter buffers around the junction. The map will also include the alternate site, shaded using a diagonal hatched pattern, and will display the river to help orient map readers to the location of the parcels.

Set the display environment

1. Click the Select Elements tool on the Tools toolbar.
2. Click the Study Area data frame on the page (the one in the upper left).

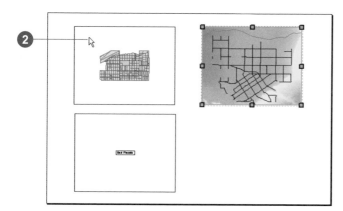

The data frame is selected and highlighted on the map, and the data frame name is shown in bold type in the table of contents. To make it easier to see the data you're working with, switch to data view.

3. Click View and click Data View.

You'll be displaying all the listed layers in this data frame, except streets, so remove the streets layer.

4. Right-click streets in the table of contents and click Remove.

The data frame should now contain the following layers, listed in this order in the table of contents:

alternate site

junction02buf

junction point

parcel02sel

river03exp

parcel01mrg

If the data frame contains any other layers, you can remove them.

Currently, alternate site, junction point, and parcel01mrg should be displayed.

Change the symbol for the parcels

You'll use the parcel01mrg layer as a background layer. This layer contains all the parcels in the study area. The foreground data in this data frame will be the layer containing only the suitable parcels (parcel02sel). It will display on top of the parcel01mrg layer.

1. Right-click parcel01mrg in the table of contents and click Zoom To Layer.

   The parcels fill the data frame.

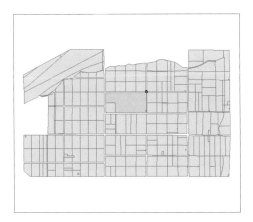

You want to emphasize the suitable parcels when you draw them on top, so change the color for the parcels in this layer to a light shade.

2. Right-click the parcel01mrg layer in the table of contents and click Properties.

3. Click the Symbology tab.

4. Click the Symbol button.

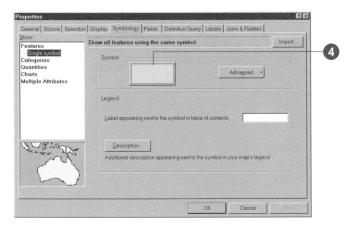

5. Click the Fill Color dropdown arrow and click Blue Gray Dust.

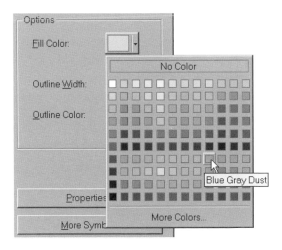

6. Click OK.

7. Click the Display tab on the Properties dialog box.

8. In the Transparent box, type 70.

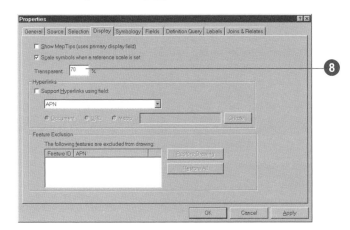

That will set the parcel fill color to be a lighter shade of the Blue Gray Dust color.

9. Click OK.

The parcels are now drawn in a light blue–gray.

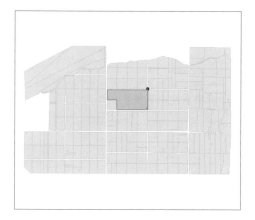

Display the suitable parcels

1. Check the box next to the parcel02sel layer to display it.

2. Click the symbol box beneath parcel02sel.

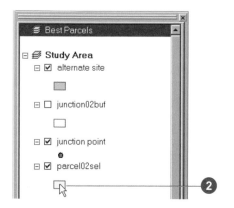

3. Click the Fill Color dropdown arrow and click Blue Gray Dust.

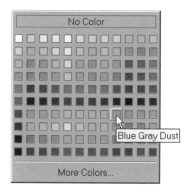

4. Click OK on the Symbol Selector dialog box.

Now the suitable parcels are drawn using a medium blue–gray, a darker shade of the color the rest of the parcels are drawn with.

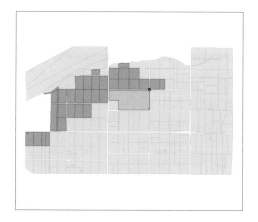

Change the symbol for the alternate site

Right now the alternate site is drawn using the default color ArcMap selected when you created the layer. You'll draw it using gray diagonal hatching so it is visible but doesn't detract from the suitable parcels.

1. Click the symbol box beneath the alternate site layer to display the Symbol Selector dialog box.

2. Scroll to the bottom and click 10% Simple hatch.

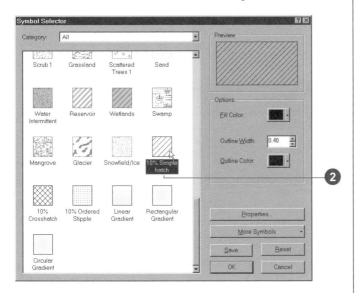

3. Click the Fill Color dropdown arrow and click Gray 40%.

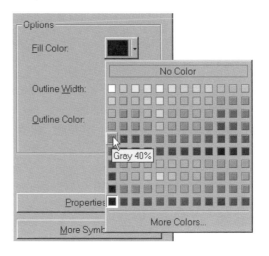

4. Click OK.

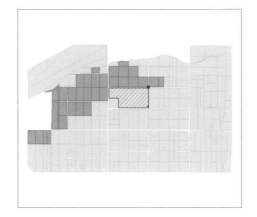

Display the river and the wastewater junction

You'll also want to display the river and wastewater junction to show where the suitable parcels are in relation to these features. You know how to change the symbology for a layer at this point, so we'll just give you the major steps. If you need help, review the specific steps in the previous sections.

Display the river03exp layer and draw it using the same symbol you used for the river in the City Overview data frame.

The wastewater junction should already be displayed. Click the point symbol underneath the layer to display the Symbol Selector dialog box. Click a symbol of your choosing. You can use the Color dropdown arrow to change the color if you like. Use the arrows in the Size text box to set the size to 14 (points).

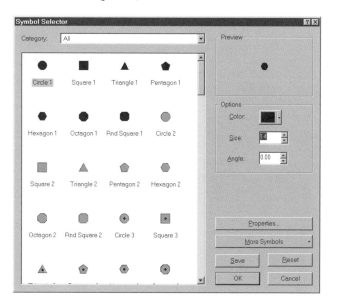

Display and label the junction buffers

Finally for this map, you'll want to display and label the 500- and 1,000-meter buffers around the wastewater junction.

1. Click the check box next to the junction02buf layer to display it.

Now you'll label the two buffers by adding text to the data frame. You can also label features using a value stored in the attribute table. You'll do that later, on the map of highly suitable parcels.

2. Click the New Text button on the Draw toolbar.

New Text

The cursor changes to a crosshair and a T.

3. Move the cursor next to the upper-right edge of the inner buffer circle and click.

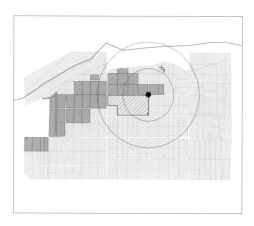

A text box appears.

4. Type "500 meters" and press Enter.

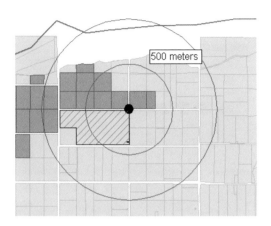

The text box is still active and highlighted with a box.

5. Click the Bold button on the Draw toolbar to make the text appear in bold type.

Bold

If you make a mistake, simply click the text to select it, press the Delete key on the keyboard, and start over.

6. With the text still selected, click and drag it to reposition it away from the outer buffer circle, if necessary.

7. Add the second label, "1000 meters", in the same manner.

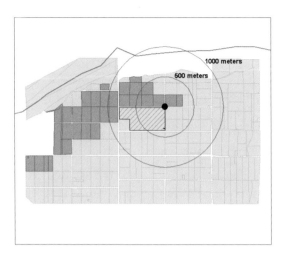

8. Click View and click Layout to switch back to layout view and display your map at this point.

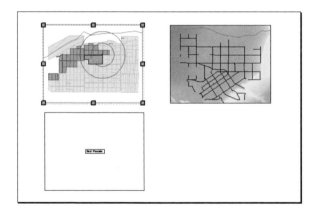

9. Click File and click Save to save your map.

You've completed two of the maps. In the next section you'll create the map of the highly suitable parcels.

Creating the map of highly suitable parcels

The third map will focus on the highly suitable parcels, color coding them by their distance from roads and the wastewater junction. You'll also label each parcel with its assessor's parcel number (APN) so people can associate the parcels on the map with the parcel report you'll create. Finally, you'll label the alternate site with its area.

You've already changed the symbols in the Study Area data frame for several of the layers, so you'll copy the layers you need from that data frame into the Best Parcels data frame. Then all you'll need to do is change the color coding for the highly suitable parcels and label them.

Copy layers from the Study Area data frame

Now you can copy the layers you need into the empty data frame. Since the changes you need to make to the layers are minimal, you can continue to work in layout view. The map will be updated as you make the changes.

The order in which you add the layers to the data frame determines the order in which they are drawn, with the most recently added drawn on top. To maintain the correct order, copy them in the reverse order to the order that they appear in the table of contents.

1. Right-click the parcel01mrg layer under the Study Area data frame in the table of contents and click Copy.

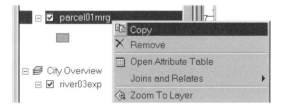

2. Right-click the Best Parcels data frame in the table of contents and click Paste Layer.

The parcel01mrg layer is added to the Best Parcels data frame and appears on the map.

Now copy the rest of the layers you need, in this order:

parcel02sel

junction point

alternate site

The Best Parcels data frame should now display these layers listed in the following order in the table of contents:

alternate site

junction point

parcel02sel

parcel01mrg

Create the layer of the highly suitable parcels

To make it easier to display and label the highly suitable parcels, you'll select them and create a new layer in the data frame. You'll do this by creating a selection expression. In Layout view, queries are performed on the selected data frame. You have been adding layers to the Best Parcels data frame, but it is not the selected data frame (Study Area is selected). Before you can query the parcels layer, you have to select the Best Parcels data frame.

1. Click the Best Parcels data frame on the virtual page to select it.

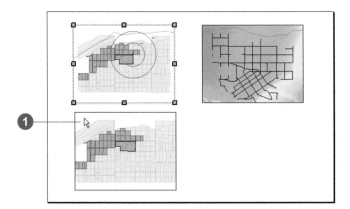

Now you can select the highly suitable parcels—those within 50 meters of a road and/or 1,000 meters of the wastewater junction. These parcels have values greater than 0 for either or both the ROAD_DIST and JUNC_DIST fields.

2. Click Selection and click Select By Attributes.

The Select By Attributes dialog box appears.

3. Click the Layer dropdown arrow and click parcel02sel.

4. Double-click ROAD_DIST in the Fields list.

5. Click the greater than button (>).

6. Double-click 0 in the Unique values list.

7. Click Or.

8. Double-click JUNC_DIST.

9. Click the greater than button (>).

10. Double-click 0.

The selection expression should look like this:

"ROAD_DIST" > 0 OR "JUNC_DIST" > 0

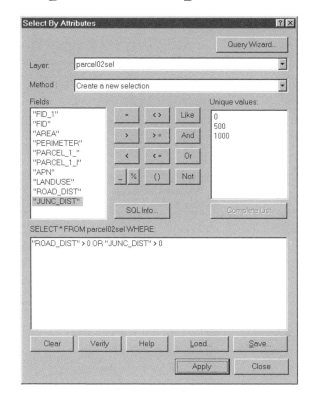

11. Click Apply, then click Close.

Now the highly suitable parcels are selected, and you can create a separate layer containing them.

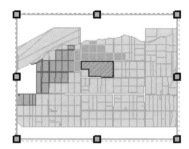

12. Right-click parcel02sel in the table of contents (under the Best Parcels data frame), point to Selection, and click Create Layer From Selected Features.

ArcMap creates a new layer in the Best Parcels data frame that contains the selected features. The default name is "parcel02sel selection". You'll rename it "highly suitable".

13. Click parcel02sel selection in the table of contents to select it, then click again to highlight the name.

14. Type "highly suitable" and press Enter.

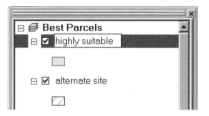

The layer is renamed in the table of contents. Now you will change the way it is symbolized and add the parcel identification numbers as labels on the map.

Color code the highly suitable parcels

You'll color code the highly suitable parcels based on both their distance from roads and from the wastewater junction, using the ROAD_DIST and JUNC_DIST fields. There are five possible pairs of values:

Less than 500 meters from the junction and less than 50 meters from a road (junc_dist = 500 and road_dist = 50)

Less than 500 meters from the junction but more than 50 meters from a road (junc_dist = 500 and road_dist = 0)

500 to 1,000 meters from the junction and less than 50 meters from a road (junc_dist = 1000 and road_dist = 50)

500 to 1,000 meters from the junction but more than 50 meters from a road (junc_dist = 1000 and road_dist = 0)

More than 1,000 meters from the junction but less than 50 meters from a road (junc_dist = 0 and road_dist = 50)

These distance values may play a part in the City Council's decision of which parcels to buy for the plant site. The parcels closer to the junction and close to a road are the most suitable, although other factors may come into play in the decision such as engineering issues (the slope and soils onsite) and economic issues (the ownership and assessed value of each parcel).

You'll symbolize the highly suitable layer so that both distance values are communicated.

1. Double-click the highly suitable layer under the Best Parcels data frame in the table of contents.

2. Click the Symbology tab.

Currently, all the parcels are drawn with a single default symbol.

3. Click Categories in the Show box and click Unique values, many fields.

The Unique values, many fields option lets you color code features based on combinations of values from up to three fields. You need only two: JUNC_DIST and ROAD_DIST.

4. Click the top dropdown arrow in the Value Fields section and click JUNC_DIST.

5. Click the second dropdown arrow in the Value Fields section and click ROAD_DIST.

6. Click Add All Values.

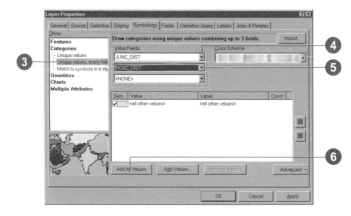

Only four pairs of values are listed. Apparently there are no parcels in the fifth category (more than 1,000 meters from the junction, but less than 50 meters from a road).

The pairs appear in this order:

500, 50

500, 0

1000, 50

1000, 0

The four pairs of values will be shown on the map with unique symbols. You'll draw the parcels within 500 meters of the junction in two shades of green (parcels near a road in a dark shade and parcels far from a road in a lighter shade) and the parcels between 500 and 1,000 meters from the junction in two shades of yellow. The remaining parcels (those meeting all the City's criteria but more than 50 meters from a road and more than 1,000 meters from the junction) will be shaded using the same blue–gray you used in the Study Area data frame.

Change the symbol colors

ArcMap uses default colors for the value combinations. You want to use two shades of green for the parcels within 500 meters of the junction and two shades of yellow for parcels between 500 and 1,000 meters of the junction.

1. Double-click the symbol box next to 500, 50.

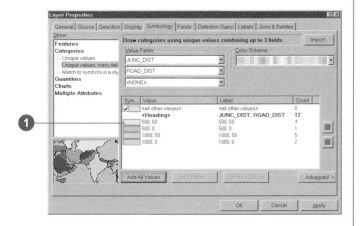

The Symbol Selector dialog box appears.

2. Click the Fill Color dropdown arrow and click Tarragon Green.

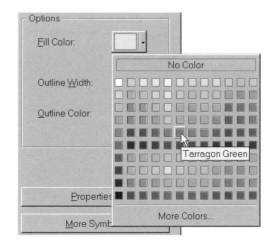

3. Click OK.

4. Now select the colors for the other values in the same manner. Use the following colors:

500, 0	Lemongrass
1000, 50	Citroen Yellow
1000, 0	Yucca Yellow

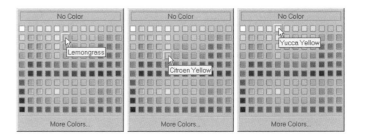

There are no other values in this case, so you can turn off the symbol for other values.

5. Click the check box next to "all other values" to uncheck it.

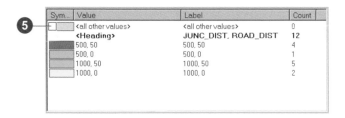

You also want to change the legend labels, so keep the dialog box open.

Change the heading and value labels

Now you'll change the labels that appear in the table of contents to make them easier to understand. These labels will also appear in the map legend when you create it.

1. Click the label field for the Heading and type "Distance to: Junction, Road". Rather than pressing Enter (which will close the dialog box), just click in the next label field to change the label.

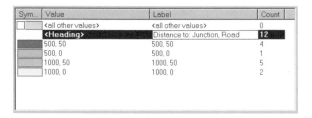

2. Click the label field for 500, 50 and type "<500m; <50m".

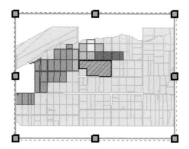

3. Change the labels for the remaining three symbols.

For 500, 0 type "<500m; >50m".

For 1000, 50 type "500-1000m; <50m".

For 1000, 0 type "500-1000m; >50m".

4. Click OK.

Now the highly suitable parcels have unique symbols based on distance from roads and the wastewater junction. The parcels shaded green are within 500 meters of the junction, and the parcels shaded yellow are between 500 and 1,000 meters from the junction. The dark-shaded parcels (both green and yellow) are within 50 meters of a road, while the lighter shaded parcels are more than 50 meters from a road.

Label the highly suitable parcels

Next, you'll label the highly suitable parcels with their APN so they can be identified in the parcel report. First, zoom to the highly suitable parcels to emphasize them on the map.

1. Click the Zoom In tool on the Tools toolbar and click and drag a rectangle around the highly suitable parcels and the alternate site.

   Since you're zooming in on the data—not the map page—make sure you're not using the Zoom In tool on the Layout toolbar.

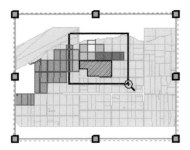

Before you add the labels, you will make sure that the correct field is used for labeling.

2. Double-click highly suitable.

3. Click the Labels tab of the Layer Properties dialog box.

4. Click the dropdown arrow to select the label field and click APN.

5. Check Label Features in this layer in the upper-left corner of the dialog box and click OK.

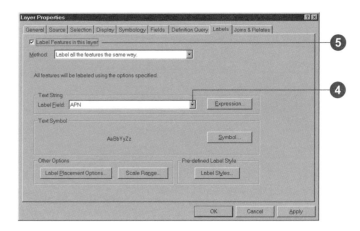

Each highly suitable parcel is now labeled with its APN.

It's hard to see how the labels will look since the map is currently scaled down to fit the screen. You can display the map at the actual size to see how it will look when printed.

6. Click the Zoom to 100% button on the Layout toolbar.

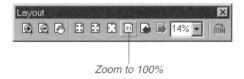

Zoom to 100%

The map is now displayed in the ArcMap window at the printed size, but you're looking at the center of the map.

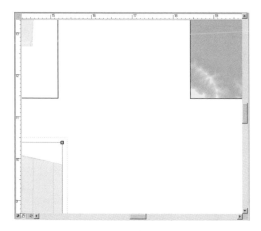

7. Click the Pan tool on the Layout toolbar and drag the map to the upper right until you can see the highly suitable parcels with their labels.

Pan

The Pan and Zoom tools on the Layout toolbar let you move around the map page, whereas Pan and Zoom on the Tools toolbar let you change the geographic extent of the data displayed in the currently selected data frame.

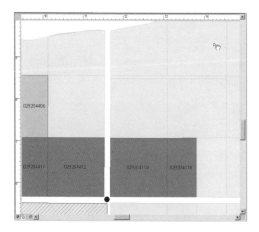

The labels look pretty good, but they could be a little larger.

8. Double-click the highly suitable layer in the table of contents to display the Layer Properties dialog box and click the Labels tab.

9. Click Symbol.

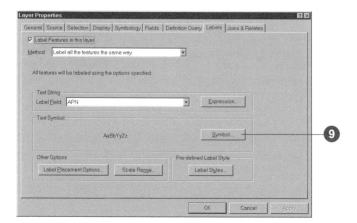

The Symbol Selector dialog box appears. The labels are currently displayed using 8-point type.

10. Click the Size dropdown arrow and click 12 to make the labels 12 point. Click OK.

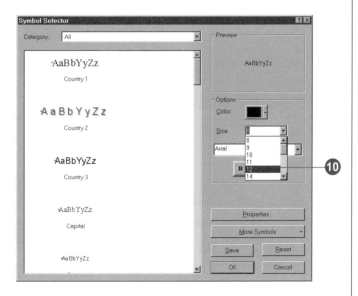

11. Click OK on the Layer Properties dialog box.

The labels are now larger and easier to read.

12. Click the Zoom Whole Page button on the Layout toolbar to see the entire map again.

Zoom Whole Page

Create the label for the alternate site

The alternate site won't be in the parcel report since it doesn't meet all the City's current criteria for a site for the wastewater treatment plant. However, you want to display its area. You'll label the parcel with its AREA field using the Layer Properties dialog box to set the properties before displaying the label.

1. Double-click alternate site in the table of contents to display the Layer Properties dialog box.

2. Click the Labels tab.

3. Click Expression.

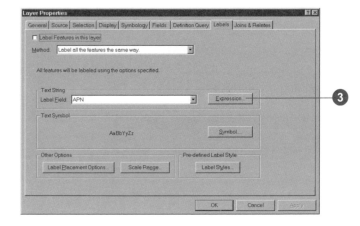

The Expression Properties dialog box appears. You can create a Visual Basic (VB) or Java™ script to customize the way labels are displayed. You'll create a simple VB script to display the area value with a suffix of "sq meters". The area value is stored in the database using several decimal places, which you don't need to display, so you'll round the value to no decimal places.

You will need to create an expression that looks like this:

Round([AREA], 0) & " " & "sq meters"

4. Click in the Expression box.

5. Type the expression in full or type all the characters except the field name, which you can drag from the Label Fields box.

The VB Round command has two arguments, enclosed in parentheses—the name of the field, AREA in this case, and the number of decimal places to round to, 0 in this case. You enclose in double quotes any text you want to display as part of the label, in this case "sq meters". The double quotes with two spaces inside will ensure there is space between the area value and the suffix when the label is displayed. You use an ampersand (&) to connect the elements of the script.

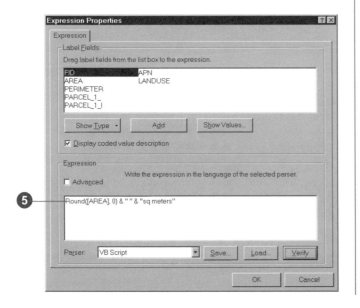

6. Click the Verify button to make sure you entered the expression correctly.

The Sample label box appears and tells you the expression is valid. It also shows you a sample of how the label will look. (Note that the value displayed in the Sample label box is not the actual value of the parcel.) If you get an error message when you click Verify, just check that you entered the expression correctly, make any changes, and click Verify again.

7. Click OK to close the Sample label box and click OK to close the Expression Properties dialog box.

Now you've specified what will be in the label. Next you'll want to change the way the label is displayed.

Change the label properties and display the label

1. Click Symbol on the Labels tab.

The Symbol Selector dialog box appears.

2. Click the Size dropdown arrow and click 12 to make the label 12-point text.

Now the label will be large enough, but it will be drawn on top of the diagonal hatching and might still be difficult to read. You'll add a mask around the text so it appears on a solid background.

3. Click Properties.

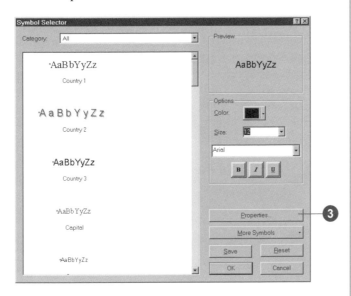

The Editor dialog box appears to let you edit the text properties. You can see that the size is shown as 12, as you just specified.

4. Click the Mask tab.

5. Click Halo in the Style panel.

The Preview panel shows you how the text will look. You want the mask to be wider to block out the hatching in the alternate site parcel.

6. Click the up arrow in the Size text box to increase the mask width to 4 points.

You also want the mask to be very light gray so it's easy to read the text.

7. Click Symbol to display the mask Symbol Selector dialog box.

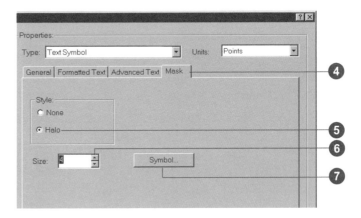

8. Click the Fill Color dropdown arrow and click Gray 10%.

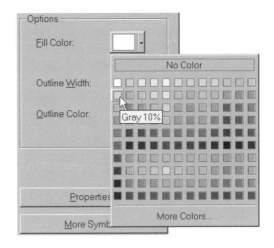

9. Click OK to close the mask Symbol Selector.

The Preview panel shows that the mask is light gray.

10. Click OK to close the Editor dialog box and click OK again to close the text Symbol Selector dialog box.

11. Check Label Features in this layer in the upper-left corner of the dialog box and click OK.

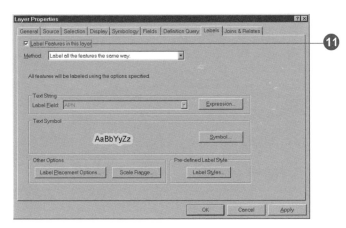

The alternate site is labeled with its area.

12. Click the Zoom In tool on the Layout toolbar and click and drag a rectangle around the alternate site.

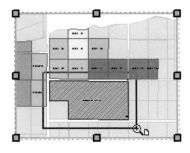

You can see the label with its mask.

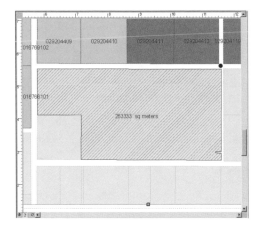

13. When you're done reviewing the label, click the Zoom Whole Page button on the Layout toolbar to view the entire map again.

14. Click File and click Save to save your map at this point.

The third map is complete, and the geographic information you need to present is now displayed in each data frame. Next, you'll create a report for the highly suitable parcels, add it to the map, and then finish laying out the map by adding the rest of the map elements.

Creating the parcel report

You'll create a tabular report of the highly suitable parcels to provide additional information about each. The report will list the assessors parcel number, area, and distance from the junction for each parcel. You'll group the parcels by distance from the junction and sort them by size.

First you'll design the report—specifying what should be included—and then you'll generate it and add it to the map.

Design the report

You'll specify the fields you want in the report, then you'll specify how to group and sort the parcels.

1. Click Tools, point to Reports, and click Create Report.

The Report Properties dialog box appears, and the Fields tab is selected. The other tabs are dimmed since you haven't yet specified any fields to include in the report.

2. Click the dropdown arrow in the Layer/Table text box and click highly suitable as the layer to create the report from.

3. Double-click JUNC_DIST to move it from the Available Fields to the Report Fields column.

4. Double-click APN and AREA to add them to the Report Fields column as well.

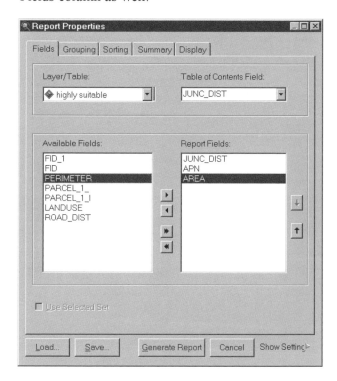

Now the other tabs are available.

5. Click the Grouping tab.

6. Double-click JUNC_DIST to specify this as the grouping field.

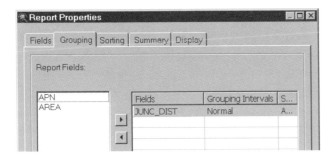

Parcels less than 500 meters from the wastewater junction will be in one section of the report, and the parcels between 500 and 1,000 meters from the junction will be in another.

7. Click the Sorting tab.

Here you specify which fields to use to sort the records and how to sort them.

8. Click in the Sort column for AREA and click Descending on the dropdown list that appears.

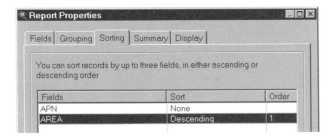

The largest parcel will be listed at the top for each group.

You'll need to change the width for the APN column to make it wide enough to display the full assessor's parcel number.

9. Click the Display tab, click Fields, and click APN.

10. Double-click in the box next to Width and type 1. Press Enter.

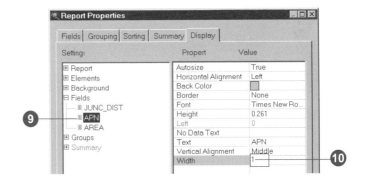

Generate the report

Now have ArcMap generate the report using the settings you have specified.

1. Click Generate Report.

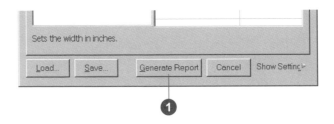

The Report Viewer appears. The viewer lets you preview the report.

The report looks good, so add it to the map.

2. Click Add.

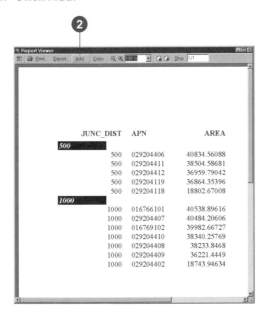

The Add to Map dialog box appears.

The report is only one page long, so you will accept the default settings.

3. Click OK.

   The report appears on the map.

4. Click the x button to close the Report Viewer and click the Close button to close the Report Properties dialog box.

   The Report tool asks if you want to save this report.

5. Click No.

6. Click the Select Elements tool, if necessary.

7. Click and drag the report into position beside the Best Parcels data frame.

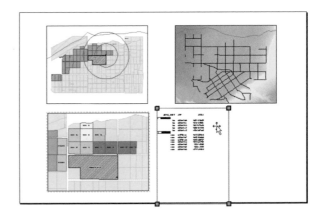

8. Click File and click Save to save your map.

Adding the list of site criteria to the map

So that the public as well as the City Council will be aware of the criteria used for finding a site for the plant, you'll add a text file listing the criteria. The City Council staff e-mailed you a text file containing a copy of the criteria. The file is stored in the project folder.

1. Click Insert and click Object.

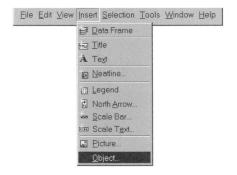

The Insert Object dialog box appears.

2. Click Create from File, then click the Browse button.

3. Navigate to the project folder, click Site Criteria.rtf, and click Open.

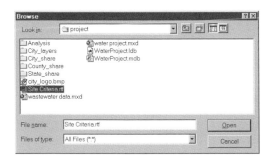

4. Click OK.

The text file is added to the map.

5. Click and drag the text to the right of the parcel report.

If you want, you can make the text smaller by using the green selection handles to resize the text block. Click the bottom-left selection handle and drag it toward the top-right selection handle (holding the Ctrl key while you drag the handle will reduce the text proportionally). It's okay if the selection box extends beyond the map page since it won't be visible once the text block is no longer selected.

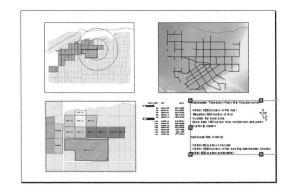

Adding the map elements

At this point you've added to the map the content you need to present to the City Council. Now you'll add the map elements to make the map easier to read and more visually attractive.

Here are the elements you will add:

- An extent rectangle to show the location of the study area on the City Overview data frame
- Map legends
- Scale bars
- A North arrow
- The map title
- The City logo
- Map reference information
- Graphic rectangles to create a title bar and to enclose the map

Add an extent rectangle to the City Overview map

You will add an extent rectangle to the City Overview data frame that will show the location of the suitable parcels relative to the rest of the City. Extent rectangles show the size, shape, and position of one data frame in another data frame.

First you'll need to select the City Overview data frame.

1. Click the Select Elements button, if necessary.

2. Click the City Overview data frame to select it.

3. Right-click on the data frame and click Properties.

   The Data Frame Properties dialog box appears.

4. Click the Extent Rectangles tab.

You will show the position of the Study Area data frame.

5. Click Study Area in the Other data frames list.

6. Click the top arrow button to move Study Area to the right list box.

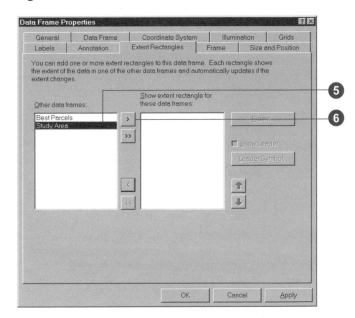

ArcMap lets you choose a variety of symbols for the extent rectangle, but the default black line will work just fine. If you want to change the symbol, click the Frame button to display the Frame Properties dialog box.

7. Click OK on the Data Frame Properties dialog box.

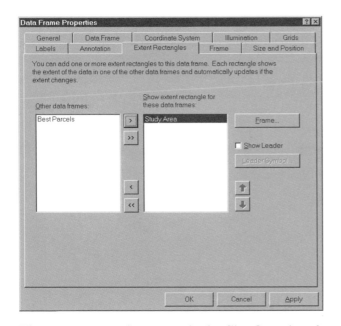

The extent rectangle appears in the City Overview data frame, showing the position and extent of the Study Area data frame, but it's clipped by the edge of the data frame.

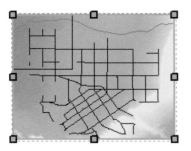

8. Click the Pan tool on the Tools toolbar and drag the layers down and to the right until the entire extent rectangle is visible.

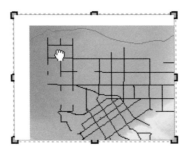

The City Council and public will now be able to see the location of the suitable parcels relative to the major streets of Greenvalley.

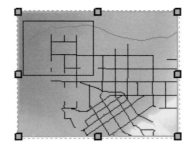

Add the map legend for the City Overview map

You'll want to add legends and scale bars for all three data frames. You'll select each data frame and create the legend and scale bar for each in turn. ArcMap automatically creates a legend based on the table of contents for each data frame. Once you've created the legend you can move, resize, and edit it.

The City Overview data frame should still be selected.

1. Click Insert and click Legend.

If the Legend Wizard dialog box appears, the ArcMap option to "Use wizards when available" is turned on (this option is under Options on the Tools menu on the Applications tab). You can proceed with the following steps using the wizard. If the Legend Wizard doesn't appear and the legend is added immediately to the map, skip to step 5 below.

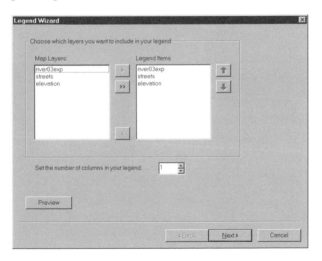

The wizard automatically lists all the layers in the data frame for inclusion in the legend. You want all the layers for this map.

2. Click Next.

You don't need a title for the legend, so double-click Legend and press the Backspace key to delete the text.

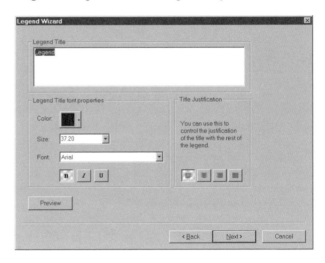

3. Click Preview.

The legend is displayed in the center of the map (you may have to move the wizard dialog box to see it). You'll use the default settings for the rest of the legend properties, so you can exit the wizard at this point.

4. Click Finish.

5. If necessary, click the selection handle in the upper-right corner of the legend box and drag it to the lower left until the legend will be small enough to fit on the page, to the right of the City Overview data frame.

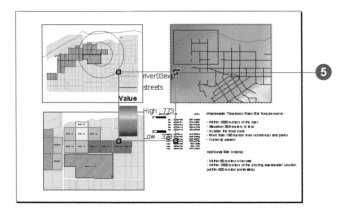

6. Click and drag the legend to the right of the City Overview data frame.

7. Click the Zoom In tool on the Layout toolbar and click and drag a rectangle around the legend so you can see it better.

You'll notice that the text for the river symbol reads river03exp, and the label for the elevation layer is "Value". ArcMap takes the legend text directly from the table of contents. You want to use more explanatory text in the legend. This is easy to change.

8. Click river03exp under the City Overview data frame on the table of contents to select it. Click again to highlight the name.

9. Type "river" and press Enter.

The legend is updated with the new text.

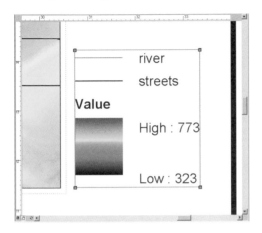

Now change the text for the elevation layer.

10. Click Value under the elevation layer to select it, then click again to highlight the name. Type "elevation" and press Enter.

11. Click the Zoom Whole Page button on the Layout toolbar to see the entire map again.

Add a scale bar for the City Overview map

Each of the three data frames on the map is drawn at a different scale, so you'll need to add a scale bar for each. You'll add a scale bar for the City Overview data frame while it's still selected.

1. Click Insert and click Scale Bar.

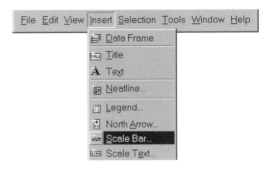

The Scale Bar Selector dialog box appears.

2. Click a scale bar that appeals to you and click OK.

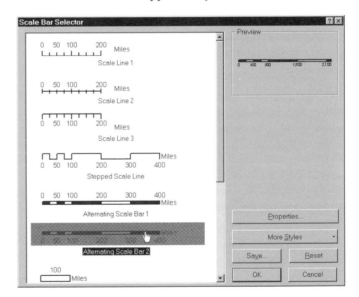

The scale bar is added to the map.

3. Click the scale bar and drag it below the City Overview data frame.

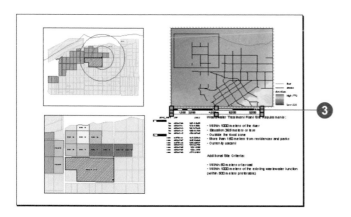

ArcMap knows the scale of each data frame and creates the scale bar accordingly.

Add the other legends and scale bars

Now you know how to add legends and scale bars to your map. We'll give you the major steps for the other two data frames, and you can work through the tasks yourself. You can review the previous sections if you need help with the specific steps.

Select the Study Area data frame by clicking it. Insert a legend. You don't need to include the junction buffers or the parcel01mrg layers in the legend, so remove them from the Legend Items list in the wizard (click each and click the < arrow at the bottom to move them off the list). You want the layers to appear in the following order in the legend: parcel02sel; alternate site; junction point; and river03exp.

Click parcel02sel in the list and click the up arrow twice to move it to the top. The layers should now be in the right order. The legend doesn't need a title, so remove it on the next screen in the wizard. Then preview the legend and click Finish to add it to the map. (If you're not using the wizard, after the legend is added to the map just right-click on it and click Properties. Then click the Items tab to change the layers that are displayed and click the Legend tab to remove the title.)

Click and drag the legend to the right of the Study Area data frame. Make the legend small enough to fit between the two data frames by clicking and dragging the upper-right selection handle, if necessary.

Change junction point to be just "junction", river03exp to be "river", and parcel02sel to be "suitable parcels".

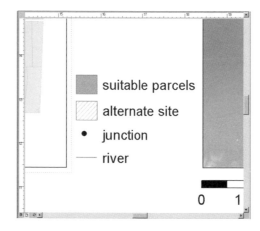

Insert a scale bar for the Study Area data frame. Use the same style you used for the City Overview and drag it under the Study Area data frame (you may want to zoom and pan on the page to position the scale bar).

(If you zoom in on the two scale bars you've added so far, you can see that the City Overview map is at about half the scale of the Study Area map.)

Select the Best Parcels data frame and insert a legend. You only need to include the highly suitable parcels, the alternate site, and the wastewater junction, in that order. This time include a title for the legend: Highly Suitable Parcels. Add the legend to the map, reduce its size if necessary by clicking and dragging one of the corner selection handles, and drag it to the right of the Best Parcels data frame, below the report.

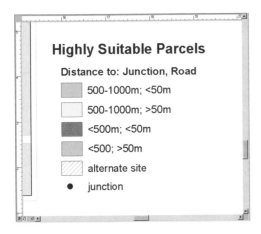

Finally, add a scale bar for the Best Parcels data frame and position it below the frame.

Save your map at this point.

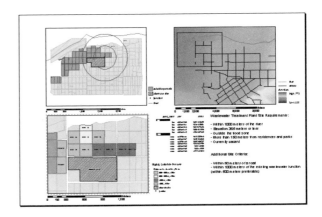

Now you are ready to add a few final map elements to finish the map. You'll include a North arrow, a map title, the City logo, and map reference information. You'll also add two graphic rectangles to tie the composition together.

Add a North arrow

You'll place a North arrow in the upper-left corner to show the orientation of the whole map.

1. Click Insert and click North Arrow.

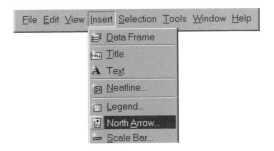

The North Arrow Selector dialog box appears.

2. Click a nice-looking North arrow. Click OK.

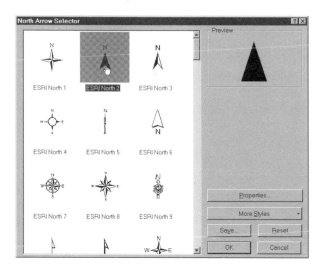

The North arrow appears on the map.

3. Click the North arrow and drag it to the upper-left corner of the map so it is a little lower than the top of the Study Area data frame.

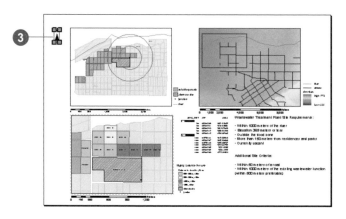

Add a map title

Next you'll add a descriptive title to the map, placed vertically along the left side of the page.

1. Click Insert and click Title.

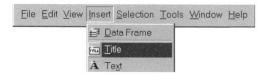

The text "water project" appears on the map. ArcMap uses the name of the map file as the default title.

2. Type "Potential Wastewater Plant Sites".

3. On the Draw toolbar, type "72" in the font size text box and press Enter.

The map title is redrawn in 72-point type.

4. Right-click on the title, point to Rotate or Flip, and click Rotate Left.

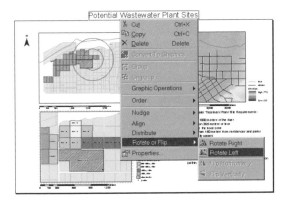

The title is rotated. Now you can place it along the left edge of the map.

5. Click on the title and drag it to the left edge of the map below the North arrow.

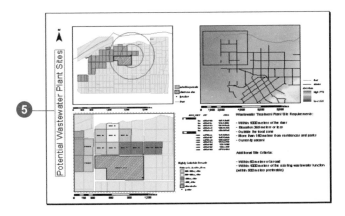

Add the City logo

The map needs to have the City logo on it. You've got a bitmap version of the logo that you've used for previous projects stored in the project folder.

1. Click Insert and click Picture.

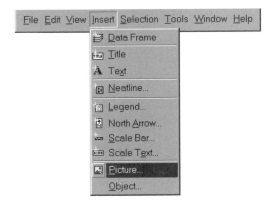

2. Navigate to the project folder.

3. Click city_logo.bmp and click Open.

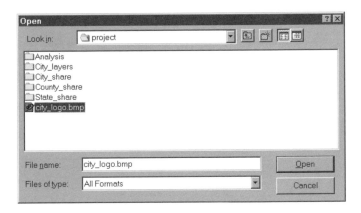

The logo is added to the map. You'll need to make it smaller and move it.

4. Right-click the logo and click Properties.

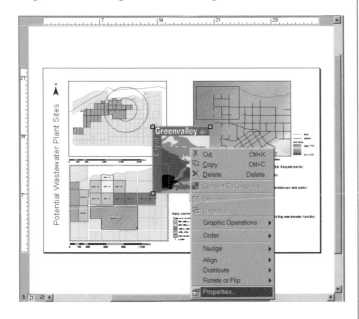

5. Click the Size and Position tab.

6. Uncheck the As Percentage box and check the Preserve Aspect Ratio box.

7. Click in the Width box and type 2.5 to make the logo 2.5 inches wide. Click OK.

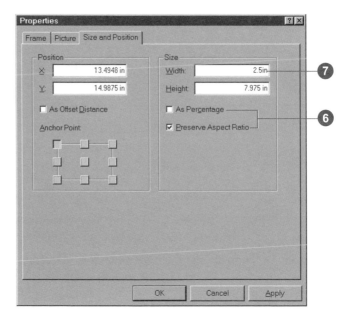

8. Click and drag the logo to the lower-right corner of the map.

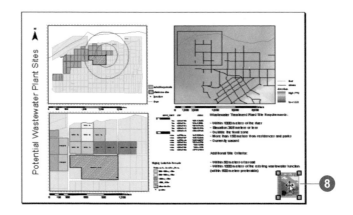

Add the map reference information

You'll want to add information about the map itself for reference. At a minimum, this should include the map projection information and the date. You can also include your name, as the author of the map, if you'd like.

1. Click the New Text tool on the Draw toolbar.

New Text

2. Click beneath the logo.

In the text box, you'll type the map projection information on the first line and the date on the second line.

3. Type "UTM Zone 11N, NAD 1983".

4. Press the Ctrl key and press Enter to add a line break (if you press Enter without the Ctrl key, the text will be added to the map immediately).

5. Type today's date on the second line.

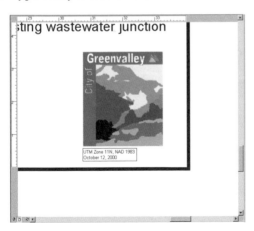

6. If you'd like, press Ctrl–Enter again and type your name.

7. Press Enter to add the text to the map.

The text box is still selected. You want to use 12-point type.

8. Click in the text size box on the Draw toolbar, type 12, and press Enter.

In an actual GIS project you might also include the sources of the data used on the map and the dates the data was collected, other contributors to the map, copyright information, and so on.

9. If necessary, click and drag the text to reposition it so it lines up with the left edge of the logo.

10. Click File and click Save to save your map at this point.

Align the map elements

At this point, you've added all the elements that you needed to the map—the extent rectangle, legends, scale bars, North arrow, title, and logo—and arranged them on the page. Before adding the graphic rectangles to complete the map, you'll want to align the data frames and other map elements.

1. Click the Study Area data frame to select it.

2. Press and hold the Shift key and click the Study Area legend, the City Overview data frame, and the City Overview legend so that all four elements are selected.

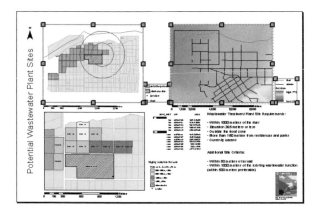

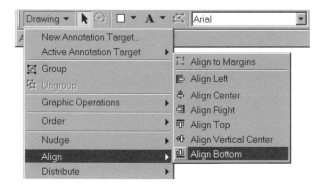

3. Click the Drawing dropdown arrow on the Draw toolbar, point to Align, and click Align Bottom.

The bottom edges of the four elements now line up. You can align other combinations of map elements the same way.

Select the scale bars below the Study Area and City Overview data frames and use Align Vertical Center.

Select the Best Parcels data frame and its legend and use Align Bottom.

Select the Study Area data frame and scale bar and the Best Parcels data frame and scale bar and use Align Left.

Select the City Overview data frame and its scale bar and use Align Left.

Select the Study Area legend, parcel report, and Highly Suitable legend and use Align Left.

You may want to double check to make sure none of the map elements overlap (use the Pan and Zoom tools on the Layout toolbar) and move any elements accordingly.

Finally, you'll add two graphic rectangles to make your map look more polished. The first will frame the title and North arrow, and the second will tie the entire composition together.

Add graphic rectangles

First you'll place a graphic rectangle behind the title and North arrow.

1. Click the New Rectangle button on the Draw toolbar.

2. Click below and to the left of the title, in line with the bottom of the Best Parcels data frame, and drag a rectangle around the title and the North arrow so the top of the rectangle is even with the top of the Study Area data frame.

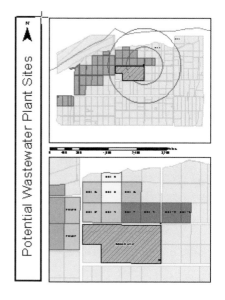

The graphic rectangle appears on the map but covers the title and North arrow.

3. Right-click the rectangle, point to Order, and click Send to Back.

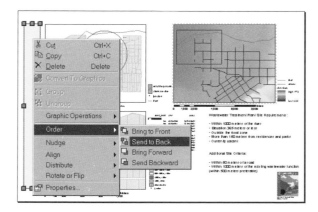

The rectangle is now behind the title and North arrow.

4. Click the dropdown arrow beside the Fill Color button on the Draw toolbar.

5. Click a light blue from the color menu.

The rectangle is drawn in light blue.

The rectangle should encompass the title and North arrow. If you need to resize the rectangle, just click one of its selection handles and drag. If you need to reposition the title or North arrow so they fit within the rectangle, just click and drag them.

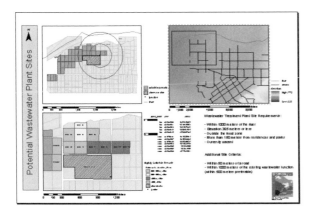

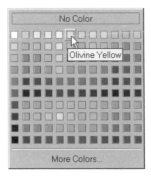

10. Click Olivine Yellow on the color menu.

Now you'll place a second graphic rectangle behind all of the elements on the page to frame the map and tie the composition together.

6. Click the Rectangle button on the Draw toolbar.

7. Click at the upper-left corner of the map and drag a rectangle to the lower-right corner of the map.

    The second graphic rectangle appears on the map.

8. Right-click the rectangle, point to Order, and click Send to Back.

    The rectangle is drawn behind the other map elements.

9. Click the dropdown arrow beside the Fill Color button on the Draw toolbar.

You have completed the poster map for the City Council meeting.

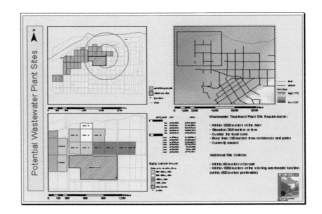

When you produce maps for publication, it is a good idea to check the final map for errors. This should include proofreading text, reviewing the symbology to make sure it is clear, and reviewing the map composition for balance. You should print the map to verify the colors—this will also make it easier to do other proofreading work.

Saving the map and printing it

Now that you have finished laying out the map, you will save a copy of it. You'll still have the draft map if you need to make more changes.

1. Click File and click Save As.

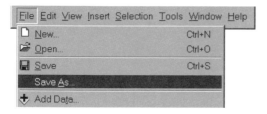

The Save As dialog box appears.

2. Navigate to the project folder.

3. Type "Wastewater Treatment Plant Sites" and click Save.

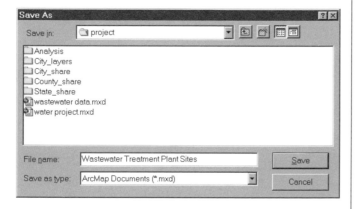

Later, when you need to view this map again, it will be available exactly as you have created it.

If you have a printer connected to your computer, you can print the map. You created the map as a D size map, so if your printer will print D size paper, you can print the map at full size; otherwise, you can scale the map down to fit your printer.

4. Click File and click Page Setup.

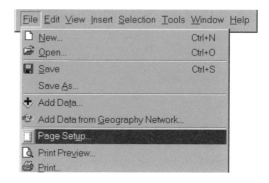

The Page Setup dialog box appears.

5. On the Printer Setup panel, click Landscape, then click OK.

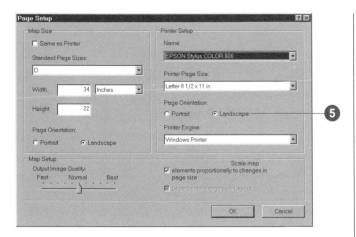

6. Click File and click Print.

7. If your printer doesn't print as large as D size, click Scale map to fit printer paper. If it does print D size or larger, skip this step.

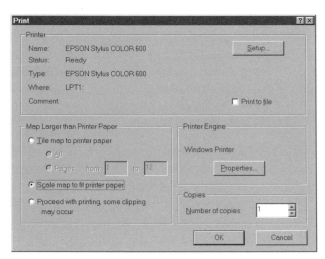

8. Click OK to print the map.

Now you have a map suitable for framing or presenting to the City Council. There were a lot of steps in this fairly complex map but, as with the analysis phase, you actually used a small set of operations—in this case, assigning symbols to features, sizing and positioning map elements, and adding text and graphic elements. Understanding how to perform these basic tasks is the key to creating just about any type of map. Of course, ArcMap gives you many more options for creating highly customized maps.

The project is finished! The City Council is likely to ask for additional analysis and a new map for their next meeting. Since the project database is complete, modifying the criteria and rerunning the GIS analysis should be relatively easy. And since you've saved a copy of your map, substituting the new analysis results will be straightforward.

You've completed a small, sample GIS project. While the scope was limited, the project process and many of the specific methods and tasks are ones that apply to a wide range of projects you'll encounter in your GIS work. The next section lists a few of the many resources that are available to you as you begin to explore the broad range of GIS applications and the specific functionality of ArcGIS.

What's next?

This book provided an introduction to using ArcGIS in a project setting. As you undertake your own GIS projects, you'll likely find you may need help with specific tasks that aren't covered in this book. You may also need to employ other ESRI software applications that weren't used for this sample project.

This section lists a few of the many resources available for learning GIS, for finding GIS data and sample maps, and for getting help. It also provides a brief overview of other ESRI software applications you may find useful.

Learning GIS

There are a number of resources available for learning about GIS and learning how to use ArcGIS, including reference books, workbooks, and courses.

Books

You can use the other books that come with ArcGIS to learn more about how to use the ArcGIS applications, how to build GIS databases, how to do GIS analysis, and how to customize ArcGIS.

When you want to quickly find out how to do a specific task, you can look it up in four reference books: *Using ArcCatalog*, *Using ArcMap*, *Editing in ArcMap*, and *Using ArcToolbox*. These books are organized around specific tasks. They provide answers in clear, concise steps with numbered graphics. Some of the chapters also contain background information if you want to find out more about the concepts behind a task. In addition, each book includes a quick-start tutorial specific to that application.

If you want task-oriented, step-by-step help creating a geodatabase, read *Building a Geodatabase*. This book will teach you how to take your geodatabase design and implement it in ArcGIS.

If your job includes designing GIS databases or developing applications, or if you want to deepen your understanding of the organization of your GIS, you'll want to read *Modeling Our World*. This book provides a broad conceptual discussion of GIS data models, with examples to illustrate the theory.

When you're ready to explore GIS analysis in more depth, read *The ESRI Guide to GIS Analysis*. This book presents the basic concepts behind geographic analysis. It also illustrates the most common methods for performing different types of analysis, using examples from a variety of GIS applications.

Exploring ArcObjects will introduce you to the development tools and environment that are available for customizing, extending, and creating extensions for ESRI enduser applications.

Self-study workbooks

ESRI Press publishes a number of self-study workbooks to help you learn specific software applications. The books consist of short conceptual overviews of specific tasks, followed by detailed exercises framed in the context of real problems. The books include a CD–ROM containing the sample data required to complete the exercises. Visit *www.esri.com/esripress* for information on currently available workbooks.

Instructor-led courses

ESRI offers over 35 different courses in various aspects of GIS, including courses in using, programming, and customizing ESRI software applications, designing and building geodatabases, and GIS management. Classes are offered at facilities throughout the United States, and internationally through ESRI distributors. For course descriptions, class schedules, and registration information, visit *www.esri.com/training*. Outside the United States, contact your local ESRI distributor for course offerings and class schedules. To find the ESRI distributor nearest you, visit *www.esri.com/international*.

Web-based courses

The ESRI Virtual Campus offers Web-based GIS courses over the Internet. The courses offer flexibility while providing hands-on experience and instructional support. To tour the Virtual Campus and for course descriptions and registration information, visit *campus.esri.com*.

Getting information on other ESRI software

There are several ESRI software applications that work in conjunction with ArcGIS to provide tools for advanced data analysis and management, including the ArcGIS extensions, ArcSDE, and ArcIMS. *What is ArcGIS?* provides an overview of the extensions and applications. You can also visit *www.esri.com/software/index.html* to get more information. Here is a brief description of each.

ArcGIS extensions

Several optional ArcGIS extensions are available for more advanced analysis and visualization of GIS data.

ArcGIS Spatial Analyst provides a broad range of spatial modeling and analysis features that allow you to create, query, map, and analyze cell-based raster data.

ArcGIS 3D Analyst™ enables you to visualize and analyze surface data in three dimensions.

ArcGIS Geostatistical Analyst lets you create a continuous surface from sparse measurements taken at sample points. In addition, Geostatistical Analyst includes tools for statistical error, threshold, and probability modeling.

ArcSDE

ArcSDE allows you to manage geographic information in your chosen DBMS and to serve your data openly to the ArcGIS Desktop and other applications. When you need a very large, multiuser database that can be edited and used simultaneously by many users, ArcSDE adds the necessary capabilities to your ArcGIS system by enabling you to manage your shared, multiuser geodatabase in a DBMS.

ArcIMS

ArcIMS™ is an Internet mapping system that provides a framework for centrally building and deploying GIS services and data. Using ArcIMS, you can deliver focused GIS applications and data to many concurrent users, both within your organization, and externally on the World Wide Web.

Finding GIS data and maps

You can complete your GIS projects more quickly and cheaply by obtaining existing GIS data, when possible. Looking at maps others have created using GIS will give you an idea of the types of projects that are possible with GIS and ways of visualizing and presenting GIS data, as well as possible sources of data for your own projects. There are a number of sources of both GIS data and maps.

GIS data

Obtaining GIS data for your project can be extremely time consuming, especially if you need to create the digital data yourself. While you will undoubtedly need to do this for some proprietary data, there is much existing GIS data available from many different sources. Base data such as streets and elevation are available from both private and public sources. In addition, local organizations are finding it increasingly useful to publish their data for others to use.

The Internet makes finding GIS data much easier than in days past. Two good places to start are ArcData℠ Online at *www.esri.com/data/online/index.html* and The Geography Network at *www.geographynetwork.com*. Both sites let you search for datasets, download free data, license commercial datasets, and create dynamic maps online.

GIS maps

A good way to see the types of projects and analysis other users and organizations are doing with GIS is to look at the maps they create. Each year at the International ESRI User Conference users display maps from recent projects they have completed. Images of many of these maps can be viewed on ESRI's Web site, at *www.esri.com/mapmuseum/index.html*. Some of these maps are also published in the annual ESRI Map Book, which is available online in ESRI's GIS Store at *www.esri.com/gisstore*.

Getting support

There are several places to get help with specific issues or questions while doing a GIS project, or to get general support when getting started with GIS. These include ESRI technical support, other GIS users, and online resources.

ESRI Technical Support

If you have an issue or question related to a specific function in ArcGIS and are unable to resolve it using the documentation or online help system, you can contact ESRI's Technical Support group for assistance. Visit the ESRI Support Web page at *support.esri.com*. The site lets you submit a request for support, search FAQs and other support documents, download utilities and updates, and communicate with other users through email lists and discussion forums.

Support services for users outside of the United States are provided by the international distributor responsible for the distribution and sales of desktop software in the user's country. To contact your distributor, visit *www.esri.com/international*.

Conferences and organizations

One of the most valuable resources you'll have as you continue working with ArcGIS is other GIS users. In addition to providing help with specific technical questions, users can offer the depth of their experience in organizational and GIS management issues. A good way to contact other GIS users in your region or in your field is through GIS conferences and industry organizations.

Each year, ESRI holds an International User Conference. The conference lets users from around the world exchange knowledge and information, get face-to-face technical support, and see the latest developments in ESRI software. For information, visit *www.esri.com/events/uc/index.html*. In addition, many local and regional ESRI user groups hold meetings and conferences several times a year. Visit *gis.esri.com/usersupport/usergroups/usergroups.cfm* for information, or contact your nearest ESRI regional office or ESRI distributor.

Online resources

There are many online resources in addition to the ones listed above. A good place to start is *www.gis.com*. This site provides general information about GIS and includes links to other resources, including professional associations related to GIS, educational resources, sources of GIS data, and many others.